The Rising Tide of Energy Consumption: Powering Development

Shiva

Table of Contents

1   Introduction 1

    1.1   Motivation 7

    1.2   Outline of Dissertation 8

2   Characterization 12

    2.1   Geological Description of Samples 15

    2.2   Laboratory Experiments 16

        2.2.1   Pore Size Distribution 16

            2.2.1.1   Mercury Porosimetry 17

            2.2.1.2   Nitrogen Gas Adsorption 21

            2.2.1.3   Comparison and Discussion 26

        2.2.2   Scanning Electron Microscope 27

            2.2.2.1   SEM Images of Horizontal Surface 28

            2.2.2.2   SEM Images of Vertical Cross Section 31

        2.2.3   Mineral Content and Composition 36

    2.3   Summary 37

3   Reconsolidation 38

    3.1   Silica Dissolution 39

        3.1.1   Experiment Series I: the Influence of pH 39

        3.1.2   Experiment Series II: the Influence of Salinity 42

        3.1.3   Experiment Series III: the Influence of Temperature 44

    3.2   Reconsolidation Experiments 48

        3.2.1   Experiment Series IV: the Mechanism of Reconsolidation 48

        3.2.2   Experiment Series V: the Key Factors of Reconsolidation 57

    3.3   Discussion 58

        3.3.1   Silica Dissolution 58

3.3.2   The Process of Reconsolidation.. 59

4   Desorption... 61

    4.1   The Description of Desorption Test.. 62

    4.1.1   Results and Analysis 63

    4.2   Desorption Processes 64

    4.3   Discussion 68

    4.3.1   Desorption Capability 68

    4.3.2   The Percent of N_2.. 70

    4.3.3   Feldspar 73

    4.3.4   Organic Matter.. 74

    4.3.5   Quartz and Clay Minerals.. 77

    4.4   Conclusion.. 78

5   Optimization 80

    5.1   Proposal and Implementation... 81

    5.1.1   Proposal... 81

    5.1.2   Implementation.. 82

    5.2   Foaming Test.. 83

    5.3   Experiment on Foam Stability.. 87

    5.4   Further Experiments... 92

6   Conclusions and Outlook.. 96

    6.1   Conclusions 96

    6.2   Outlook... 97

References... 99

Appendix A.. 104

Appendix B 107

Appendix C 128

1 Introduction

As one of the hottest topics over the world, energy consumption has always been connected with the economic development. The statistical data from British Petroleum (BP) and World Bank show that the gross world product (GWP) is proportional to the gross energy consumption (Agency, 2009; Wei & Liao, 2016). From 1980 to 2009, the average correlation coefficient between the two variables was approximately 0.995. In this period, GWP increased from 17.8 to 39.4 trillion USD (at constant prices), corresponding to an annual growth rate of 2.8%; per capita GWP increased from 4000 to 5800 USD, which returns an annual growth rate of 1.3%; the worldwide energy consumption increased from 6.6 billion tonne of oil equivalent (toe) to 11.2 billion toe, indicative of an annual growth rate of 1.8%; the energy consumption per unit of GWP decreased from 3.73 to 2.84 toe/104 USD with a decrease rate of 23%; and the energy consumption per capita increased from 1.49 to 1.65 toe, which gives an increase rate of 11%. Globally, while the world economy grew by 1%, the world energy demand increased by 0.64% (Agency, 2009).

As an economic downturn is undesired, the energy consumption has been increasing year after year around the world in both developed and developing countries. According to the research of BP, the global energy trading has kept a quickly increasing trend since the 1990s with the globalization and development of transport technology (Global, 2010). The global oil trade increased from 1.34 billion tons in 1990 to 2.54 billion tons in 2008, with the ratio of oil trade to oil consumption increasing from 47 to 64% (Wei & Liao, 2016). Plotted in Figure 1.1, the latest data show that global energy consumption increased by 2.9% in 2018. The growth was the strongest since 2010 and is almost double of the average increase within recent years.

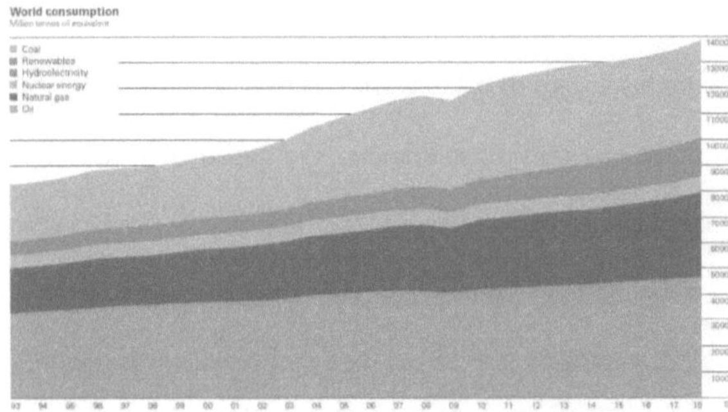

Figure 1. 1: World consumption (Dudley, 2018)

More interestingly, the data on energy consumption indicated not only the increasing tendency but also the change in the energy consumption structure. Growth was particularly strong in the case of gas (168 million toe, accounting for 43% of the global increase). Although the reasons have been differently reported, there are a few convincing perspectives which led to the adjustment of consumption structure.

The environmental issues are on the top. The over exploitation and combustion of fossil fuels to meet the energy needs are posing serious threats to the environment and contributing to environmental pollution, ecological damage, and global warming (Wei & Liao, 2016). Compared to the petroleum, combustion of gas is more friendly to the environment. Secondly, energy shortage is another threat for all the countries in the process of industrialization and urbanization. When the supply of oil cannot meet the demand of consumption, it is an opportunity for rising the proportion of natural gas.

However, it seems that the conflict between demand and consumption is impossible to be solved in a few years. Since there is no way to decrease the consumption of energy, an optimization of the consumption structure seems feasible to alleviate the problem. This optimizing solution has been conducting in America and proved to be effective. In the USA, the gas consumption has become higher than that of petroleum since 2010. According to the data released by BP, energy consumption in the United States was 2.18 billion standard toe, corresponding to 19.5% of world aggregate consumption. Among all energy sources, petroleum consumption was 0.84 billion tons (21.7%), coal consumption was 0.50 billion tons (15.22%) and natural gas consumption was 590 billion kiloliters (22.2%) (Global, 2010). These statistics predicted the tendency of gas utilization to replace petroleum, which ruled American market in the past.

Obviously, the adjustment and optimization of energy structure brought many benefits. One of them is the decreasing CO_2 emission. With the technological progress and optimization of used energy, most countries have shown a decreasing trend of emission. Another benefit is the change of energy consumption per unit of gross domestic product (GDP). From 1980 to 2008, energy consumption per unit of GDP in the United States decreased by 44% and that in Japan decreased by 23%. For the developing countries, it decreased by 19% in India and by 66% in China (Agency, 2009; Global, 2010; Wei & Liao, 2016). The sharp drop in China played a positive role in decreasing global energy demand and greenhouse gas emissions (Wei & Liao, 2016).

As the world's main energy consumer and the largest developing country in the world, the situation in China is completely different and more complicated.

The first difference between China and other countries is that China relies on coal heavily. Characterized as a high carbon, high pollution, and high emission source, coal represented a dominant constituent of the total energy consumption in China (61.8%), while oil and gas accounted for 18.3% and 6.4%, respectively. Consequently, the overusing of coal results in high carbon emission.

The second difference is the technologies to explore the conventional formation, for instance, oil reservoirs. Compared to industrial countries, technologies of exploration, drilling, fracturing, and production in China remain far behind. The directly result is the low yielding which escalated the energy shortage. Furthermore, the cost for reserves to be exploited is also many times higher than other lands. Except for the high pollution and relatively undeveloped technologies, the amount of conventional oil reservoirs is reducing as well.

Under this background, the proportion of gas demand to total energy is rising sharply in recent years. According to the statistics of the international energy agency (IEA), gas is the second energy supply. In 2017 the total supply of gas in China achieved 3,106,799 ktoe. The Opinions on Accelerating the Use of Natural Gas, promulgated by the Chinese government, also clarified that China will gradually promote gas as one of its main clean energy sources. The China National Energy Administration highlighted gas as an important basis for China's realization of a low carbon development.

To meet the energy demand, and to decrease the emission of greenhouse gas, the adjustment of energy consumption from coal to gas is the best alternative for China in this stage. However, this change is not easy to achieve. The total production cannot meet the demand, even though China's natural gas production grew at an annual average rate of 15% from 2000 to 2007. The rest of gas consumption relies on import. With a leading position in the list of imported energy, China has imported almost 20 billion cubic meters gas in 2018. Moreover, a prediction interpreted the challenge for China in the future. The import will rise year after year from 2018 to 2040. Data explain that China will consume

400 bcm in 2040, about 150 bcm in the case of domestic and 250 bcm in the case of import.

According to data provided by the China National Petroleum Corporation (CNPC) based on a national survey published in 2005, China's natural gas resources amounted to 56 trillion cubic meters (tcm) prospectively and 35 tcm geologically. The country's recoverable resources were estimated at 22 tcm, which have increased by 70% compared to the previous survey conducted in 1994. At the end of 2008, CNPC announced that China's total proven reserves amounted to 5.94 tcm, including 3.09 tcm of technically and economically producible reserves (Nobuyuki, 2009). In 2007, Cedigaz estimated that China's proven reserves amounted to 3.7 tcm, while the IEA estimated China's recoverable, proven and probable reserves from identified fields to amount to around 5.0 tcm. In addition to conventional gas resources, the 2005 survey identified the huge potential of coal-bed methane (CBM) as constituting 37 tcm of China's geological resources and 134.3 bcm of proven reserves, despite the fact that the current production level amounts to only 0.38 bcm per year. The question is then, since China is one of the countries which have great natural gas reserves, why it still needs such a huge amount of import.

CNPC detains a share of about 75% in domestic natural gas production and is accelerating the development of major gas fields with the aim of increasing production. According to CNPC, the dominant producer in China, as a result of the rapid increase in production in recent years, the reserve to production ratio declined from 51 in 2002 to 47 in 2007 despite the increase in the discovery of reserves. However, CNCP has found that the extraction of upstream gas resources has become increasingly difficult and complex due to the substantial depth of these basins as well as their high pressure, loose sandstone, volcanic pools and high sulfur content. At this moment China has consistently focused on energy shortage, as the worsening shortage in natural gas has brought new challenges.

In the meanwhile of innovating technology for conventional gas exploration, to explore the unconventional reservoir is plan B. The term "unconventional gas resources" refers to natural gas from coal (also known as coal-bed methane, CBM), tight gas sands and gas shales (Geny, 2010). According to data from the Ministry of Land and Resources, at the end of 2013, China had 11, 12 and 25 trillion cubic meters of remaining technically recoverable resources of coalbed methane, tight gas and shale gas, respectively, which are still in the early stage of development (Hou, Xie, Zhou, Were, & Kolditz, 2015).

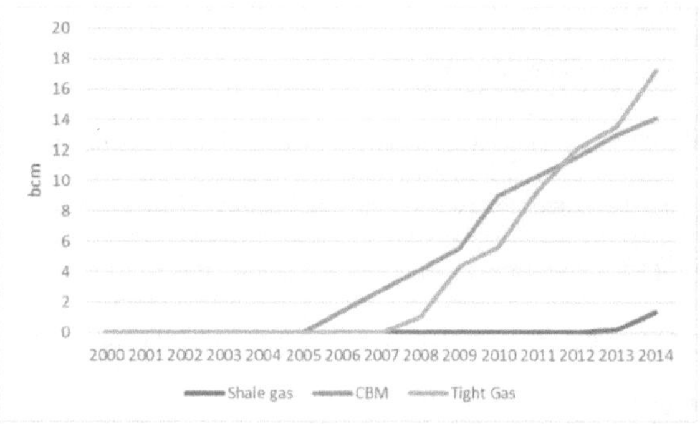

Figure 1. 2: The production of unconventional gas in China (IEA)

IEA describes the rapid development of unconventional gas. Statistics presented in Figure 1.2 show that from 2006 the production of CBM and tight gas gradually increase year after year. In 2014, according to IEA, the yielding of CBM was more than 14 bcm. Compared to the CBM, tight gas increase more rapidly, from 1 bcm in the beginning to 17 bcm in 2014. In contrast, the development in shale gas seems stagnant. Until 2014, the supply from unconventional gas was negligible, especially the production in shale layer. Although there was negligible production in this stage, it does not mean development of shale gas is worthless. Shale gas is more environmentally friendly and attractive compared to other energy resources due to its ecological advantages (low levels of carbon dioxide emission) and safety qualities (insignificant sulfur dioxide and H_2S contents) (B. Bai, Elgmati, Zhang, & Wei, 2013).

Having the largest shale gas reserve in the world, China is vigorously developing its shale gas resources. According to the results released by the Ministry of Land and Resources of China, the country has 134 and 25 trillion cubic meters of shale gas geological reserves and technically recoverable reserves, respectively (Xi Li et al., 2020). The National Energy Administration (NEA) of China forecasts that China's shale gas production will exceed 30 bcm in 2020 (Haifang, 2017). In 2018, China's shale gas production merely exceeded 10 bcm, reported by CNPC. Seen as an unconventional alternative to natural gas, shale gas is expected to become an important direction of energy selection in the future in China. The related Chinese policies show that vigorously developing and effectively using shale gas to ensure an efficient energy supply make up a long term energy strategy for meeting the demand for natural gas.

However, current shale gas development is stagnant in China due to challenges. The most important mission in this moment is to solve the problem of productivity impairment.

1.1 Motivation

As a type of unconventional reservoir, shale formations consist of regions where clay or mud minerals and fine-grained quartz have accumulated layers upon layers over millions of years (Fjar, Holt, Raaen, Risnes, & Horsrud, 2008; James G. Speight, 2013a, 2016). Because of sedimentary deposition, overburden pressure has been built up and firmed up the formation into an impermeable rock formation (Fjar et al., 2008). Completely different geology makes the great difference to conventional reservoirs and difficulty in the extracting of shale gas. What's positive is that extracting shale gas from shale gas reservoirs by using hydraulic fracturing and hydro fracking methods is now proven feasible from numerous operations in various shale gas reservoirs in North America (James G. Speight, 2016).

To successfully fracture and extract the gas trapped within the unconventional shale layer, one key characteristic is the breakability of the shale, which is strongly depending on the mineral composition. A high siliceous and calcareous content, combined with a low amount of clay or mud (less than 30% for commercial production), increases the shale's ability to fracture (Fjar et al., 2008). This is contrary to the deformation of the reservoir rock when pressure is applied, which has raised major problems in production (Bell, Kramer, Zajanc, & Aspittle, 2008), for instance, the reconsolidation of fractures, and the disappearance of proppants.

Another critical element related to the relatively complicated geology of a shale formation is that the source rock is most often the reservoir rock as well, meaning that the resource like shale oil and gas has not migrated from where it was deposited and matured (Fjar et al., 2008). This is because of the low level of permeability, most often in the range of 0.001-0.1mD where the hydrocarbons are only allowed to escape when natural and/or mechanical fracturing occurs (B. Bai et al., 2013; El-Sayed, El Domiaty, I Mourad, & Lotfy, 2018; James G. Speight, 2013b).

To increase the productivity of shale gas, the causes of productivity impairment must be analyzed before the theoretically feasible methods are presented in this thesis.

1.2 Outline of Dissertation

As geological and geochemical evaluations are the most basic requirements in the exploration and development of shale gas reservoirs projects, therefore this work will cover the essential information for Yanchang Formation characterization. The processes of dissolution and reconsolidation, which are very important in regard to the long-term productivity and they will be investigated in this work considering different factors.

Based on the statistics from the well logging evidence of exchange between N_2 and CH_4 could be established. Factors effecting the gas entrapment in shale matrix will be studied under consideration of mineralogy, reservoir conditions and organic matter.

The main part of this study is focusing on the fracturing fluid, N_2-foam generated by adding anionic surfactant as well as few other essential additives, e.g. polymers. The effectivity of such fluid has been investigated. The role of this fracturing fluid in case of CH_4-extraction will be explained.

The presentation in this dissertation follows the rules of cause and effect. It is worthwhile to mention that most statistics in Chapter 2 has been published in previous PhD dissertation of Changliang Fang, *Laboratory and Modeling Study of Fracability in Lacustrine Gas Shale of Ordos Basin, China, 2017*. Although most investigations have been shared with this dissertation, the interpretation has been reorganized independently to establish a background for further investigations. The purposes of these figures are different from previous study, but relevant for this work.

In addition to intensive literature review, more sources can be considered to find factors and approaches to optimize fracture fluid. Chapter 3 and 4 highlighted the key findings to figure out the research direction in this work.

Chapter 1 starts with a general overview of energy consumption structure followed by an introduction of the adjustment of structure. The focus is then laid on the current issues related to energy shortage in China. Although China has made the efforts to optimize energy consumption and increase the proportion of natural gas, the domestic production of natural gas cannot meet the demand. Unconventional gas has been regarded as an alternative energy, especially the shale gas which has been reported frequently with a great reserve in China. A chance brings along new challenges. How to increase the productivity of shale gas is worth to be studied.

Chapter 2 focuses on the investigation of characterization of shale matrix and in formation. The laboratory experiments have been conducted. Through nitrogen gas adsorption and mercury porosimetry, which are suitable for different precision, the pore size distribution has been interpreted. With the help of scanning electron microscope (SEM), the micro structure of the sample was presented, which is meaningful to the implement of fracturing. In the term of composition which has an influence on desorption, the X-Ray shown the content of minerals. Although it is a basic element, it shows the solution to increase production.

Chapter 3 contains a comprehensive investigation of reconsolidation. Firstly, based on the results of Chapter 2, the factors which have influence on dissolution of silica in formation had been studied, specifically quartz. And temperature played a great role in dissolution. Alkaline condition seemed suitable for further dissolving. The phenomenon of reconsolidation has been observed in Series IV. The further details had been revealed in Series V. Both temperature and confining pressure were necessary for explaining the mechanism of reconsolidation.

Having been interpreted in Chapter 2, further details about the characterization of shale are discussed in Chapter 4. What different is that in this section the researched object is the whole formation, and the relationships between desorption capacity and different component have been analyzed. As a result, the content of quartz showed positive relationship with storage. Furthermore, the feldspar serviced space for absorbed gas, even though, it is hard for the feldspar to trap gas in the matrix. The most interesting finding was the exchange between N_2 and CH_4, which would benefit the exploration of production.

Chapter 5 described how to optimize the fracturing fluid for increasing production. Based on the theory of exchange between N_2 and CH_4 existing in the formation, a possible suggestion is to risen the content of nitrogen in the fluid. In addition, to extend the exchange time also seems like another good supplement. Considering that less water could effectively prevent reconsolidation, the foam fracturing fluid may be an optimal choice. In this section, ABS and K_{12} were implemented as foam agents. For more uniform size and longer half-life period, seven more different additives were mixed with two foam agents. Finally, three formulas have been recommended for further research. Last Chapter summarized this thesis. Several experiments have been concluded. Also, disadvantages have been listed to point the further direction.

2 Characterization

Shale formations are the most abundant sedimentary rocks in the Earth's crust. These geological rocks are rich in clay, typically derived from fine sediments, deposited in fairly quiet environments at the bottom of seas and lakes, having then been buried over the course of millions of years (James G. Speight, 2013a). After long time sedimentation and tectonic movements, quartz, calcite, and clay have constituted these special sedimentary rocks with ultra-low permeability, which results in different geological functions in various fields. In petroleum geology, organic shale formations are sealed rocks that trap natural gas (James G Speight, 2014). In reservoir engineering, shale formations are flow barriers. During drilling, the bit often encounters greater shale volumes than reservoir sands (James G. Speight, 2013a).

Prior to the 1980s, the formation with the low permeability was not a target of exploratory efforts (Montgomery, Jarvie, Bowker, & Pollastro, 2005). And initial recoveries from these reservoirs has been believed largely uneconomical. Nevertheless, substantial changes have occurred in the past decades in the natural gas industry. Specifically, there has been rapid development of technologies allowing the recovery of natural gas from shale formations (James G. Speight, 2013c).

With the revolutionary technologies, for instance, horizontal drilling and artificial stimulation, there was a great successful shale gas revolution in the USA. After that, the actions to explore and extract shale gas have been conducted around the world. Unfortunately, few countries have shared their good breaking news with others. In order to figure the differences out, analyzing has been started from the geology.

The obvious difference with outer formations is that shale formations can serve as pressure barriers in basins, as top seals, and also as reservoirs of shale gas because of the ultra-low permeability (James G. Speight, 2013a). It means that low permeability shale formations are not only helpful for production of other reservoirs, but also could be the source of production, which makes the conventional technologies to disagree with the shale reservoirs.

It is necessary to stimulate the reservoir by creating a fracture network to give enough surface area, which in turn allows sufficient production from the additional enhanced reservoir permeability (James G. Speight, 2013c). Through hydraulic fracturing, i.e. hydro-fracking methods, natural gas production from shale gas reservoirs was proven feasible from numerous operations in various shale gas reservoirs in North America (Kundert & Mullen, 2009). Effective and significant technologies to increase commercial production by enhancing the permeability have been spread among regions.

The related concepts and technologies are now relatively mature in the USA, and lessons can be drawn from the experience of similar basins worldwide (S. Chen et al., 2011; Montgomery et al., 2005). In addition, the technology was primarily developed in the Texas Barnett Shale and applied to other shale lay resources, often with a one-method-fits-all approach (James G. Speight, 2013c).

However, geological conditions are ideal and probably unrepeatable. In other words, each gas shale basin is different and each has a unique set of exploration criteria and operational challenges. It is difficult to arbitrarily apply a single genetic model of productive shale gas from the USA to China. The uncertainties and difference of reservoir properties and fracture parameters have a significant effect on shale gas production in

China, making the process of optimization of hydraulic fracturing treatment designed for economic gas production much more complex.

Experience shows that continued geologic and engineering analysis as well as coupled more effective completion techniques progressively increase well performance and encouraged interest from other operators. There is now a realization that the Barnett Shale technology needs to be adapted to other shale gas resources in a scientifically technologically structured manner (James G. Speight, 2013c).

Furthermore, according to Kundert and Mullen, what needs to demonstrate firstly is that the thorough characterization of shale gas matrix and the formation (Kundert & Mullen, 2009). It is extremely important to identify reasonable ranges for these uncertainty parameters of unconventional reservoir and evaluate their effects on well performance (James G. Speight, 2013c). In addition, maximization of reservoir production ability can only be achieved by a thorough understanding of the occurrence and properties of the shale gas resources as well as the ability of the gas production from the reservoir (Kundert & Mullen, 2009).

To summarize, geological and geochemical evaluations are the most basic tasks in the exploration and development of shale gas reservoirs projects. The combination of geological framework, geochemistry, and field characteristics also has a great influence on the development of shale gas (Kundert & Mullen, 2009; Montgomery et al., 2005).

Geological analysis has been identified and, to a certain extent, has characterized the reservoir portions of shale. Geochemical data have proven essential for explaining shale potential and observed patterns of productivity (S. Chen et al., 2011).

In order to figure out the causes of low yielding and maximize the production, the basic characterization of the specific shale formation should be initiated with researches on bearing rocks. It relates to hydrocarbon reserves, mechanical properties and design of production (Fang, 2017; Fang & Amro, 2014; Fang, Amro, Jiang, & Lu, 2016; Y. Liu & Moh'd M, 2018). This chapter focuses on the laboratory experiments for investigating the porosity characters of the shale core samples from Mesozoic Triassic Yanchang Formation from Ordos Basin, northwestern China.

2.1 Geological Description of Samples

The investigated shale samples in this section, as shown in Figure 2.1, are cored in the Chang 7 member of Mesozoic Triassic Yanchang Formation in Ordos Basin of China. The well location of these shale samples is showed in Figure 2.2.

As one of considerable oil-bearing basins in China, Ordos Basin, located in the western part of North China platform, is a large-scale superimposed intracratonic basin formed after multi-periodical tectonic activities (Fang & Amro, 2014). It is generally believed that shallow deposit of sedimentary strata in China is continental sediments and deep deposit is marine sediments, which is opposite to the situation in Middle East, United States, Russia and Europe. In the Late Paleozoic-Middle Triassic period, Ordos Basin is in the process of transforming from marine and continental alternately sedimentary environment to continental sedimentary environment, as Wu described.

Figure 2. 1: Shale sample of Yanchang Formation

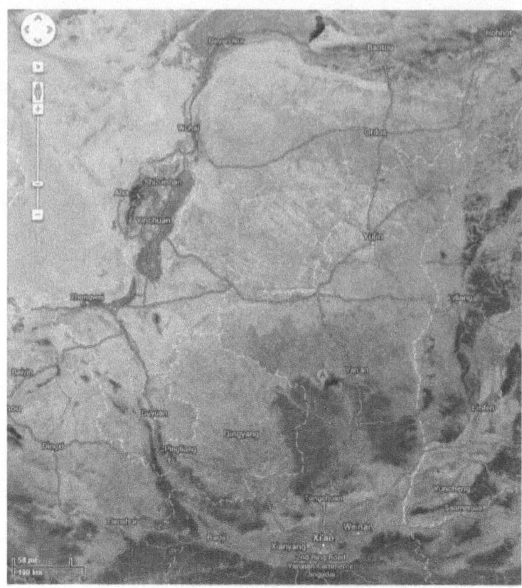

Figure 2. 2: The well location of shale sample

Mesozoic Triassic Yanchang Formation is a terrigenous clastic rock sedimentary system mainly with inland lakes-delta deposition. The area of Yanchang Formation is generally divided into 10 lithological combination members (Fang, 2017; Fang & Amro, 2014; Fang et al., 2016). The core samples tested in this chapter are from Chang 7 member of Mesozoic Yanchang Formation. This section of Yanchang Formation is the main oil source rock of lacustrine deposit in Ordos Basin, and the lithology of Chang 7 member is mainly dark mudstone, oil shale with thin layer of powder, fine sandstone, meanwhile pyrite particles are produced (Fang, 2017; Fang et al., 2016).

2.2 Laboratory Experiments

2.2.1 Pore Size Distribution

As well-known, all of the porous material has number of pores with various sizes. In terms of shale, the pore size distribution is related to hydrocarbon reserves, mechanical properties and design of simulation. It is meaningful to measure this property at the beginning. Following two sections present the mercury porosimetry and nitrogen gas adsorption of shale sample from Ordos Basin, China.

2.2.1.1   Mercury Porosimetry

Mercury porosimetry is widely used in the porosity and pore size distribution measurement, especially in rock geological analyzing. The surface tension of mercury is so great that it cannot get into the pores without external force. Assuming the volume of mercury cannot vary according to changing pressure, the pore volume equals the mercury injection volume. Washburn interpreted an equation (Washburn, 1921) to show the relationship between pore diameter and pressure in porous material:

$$PD = -4\gamma \cos \theta$$

Equation 2. 1

Where,

P-pressure of external force;

D-pore diameter;

γ-surface tension of mercury;

θ-contact angle between mercury and the solid surface.

When surface tension γ and contact angle θ are fixed, pore diameter D can be calculated with known pressure P. This Equation 2.1 also shows mercury can get into smaller pores with the increment of pressure. According to the description of Sing and Washburn, this clue indicates that pore size distribution could be presented by pore volume distribution and expressed as (Fang & Amro, 2014; Fang et al., 2016; Sing et al., 1985; Washburn, 1921):

$$f(D) = \frac{dV_c}{d(\log D)}$$

Equation 2. 2

Where,

$f(D)$-function of pore diameter distribution;

V_c-cumulative pore volume;

D-pore diameter.

Thus, the function including $f(D)$ and D shows the pore size distribution curve.

In mercury intrusion test, the parameters, such as pressure, pore diameter, intrusion volume, and surface area, are recorded automatically. The pressure increased from 0 to 200 MPa. While intruded volume increased and total value without compressibility, correction is 5.36 mm^3/g at the maximum pressure (200 MPa). Then approximate 2.5 mm^3/g volume of mercury discharged when pressure decreased from top pressure in desorption process as Figure 2.3 shows.

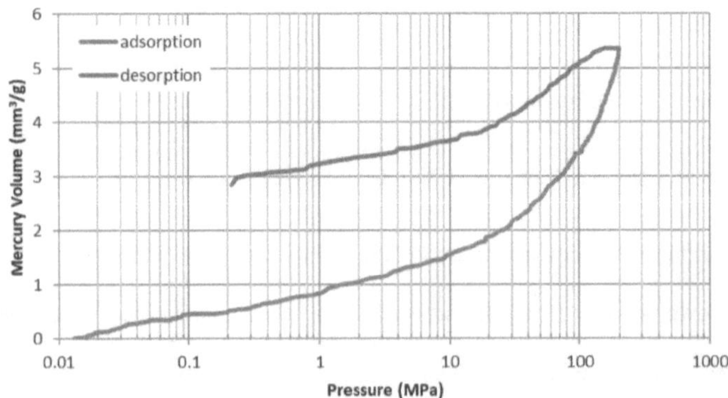

Figure 2. 3: The process of pore size distribution measurement using mercury porosimetry (Fang, 2017)

The result of mercury porosimetry showed that the porosity of this shale sample is 1.35%, which is much smaller than common shale. Meanwhile, the average pore diameter is

20.01 nm and median pore diameter of 27.38 nm. Figure 2.4 shows that log cumulative pore volume is partly linear proportional to log pore diameter. It also shows that pores with diameter below 5 µm compose about 90% of the total pore volume. According to the pores classification by International Union of Pure and Applied Chemistry (IUPAC), mesopores and macropores have contributed 60% and 40%, respectively. It should be noted that measured pore diameter starts at 7.33 nm, because 200 MPa pressure is not high enough for mercury to access in pores smaller than 7 nm diameter.

The log specific surface area curve, in analogy to that of cumulative pore volume curve, has partly linear relation with log pore diameter, as indicated in Figure 2.5. The specific surface area with pore diameter above 100 nm is almost two orders of magnitude smaller than that with pore diameter below 100 nm. This means that specific surface area almost always increases when pore diameter is smaller than 100 nm. And the smaller the pore diameter, the higher the specific surface area increment. It means that pores mainly exist with diameters less than 100 nm. In the range from 10 nm to 100 nm, smaller pores are dominant over larger pores.

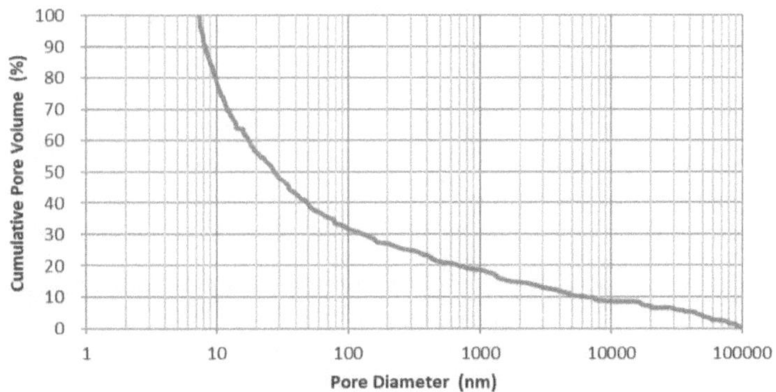

Figure 2. 4: Cumulative pore volume curve (Fang, 2017)

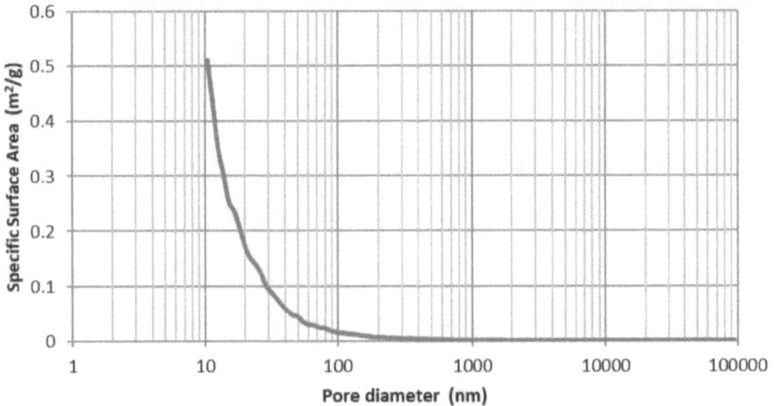

Figure 2. 5: Specific surface area curve (Fang, 2017)

The pore volume distribution of shale samples is showed on Figure 2.6. The pore

diameter ranges from 7.33 nm to about 100 μm, because maximum pressure of this

mercury intrusion test (200 MPa) is merely able to push mercury into minimum pore with

about 7 nm diameter.

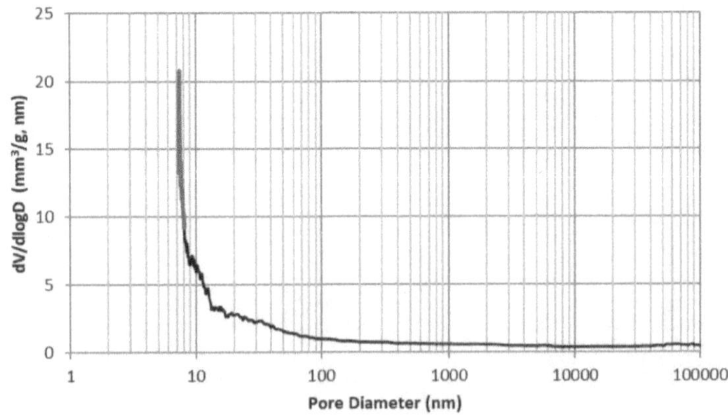

Figure 2. 6: Pore size distribution curve (Fang, 2017)

However, it is believed that the result under the highest pressure has large errors. In Figure 2.6, the pore size distribution curve has irregular sharp increment and falls in the end of minimum pore diameter (red line), which is considered incorrect. After mercury porosimetry with more than 400 MPa pressure in shales, Bustin pointed out that at very high mercury injection pressures (Bustin, Bustin, Cui, Ross, & Pathi, 2008), compressibility of the rock structure, possibility of breaking the particles, and opening closed pores will affect the data results. Hence, with the increment of pressure the mercury access into the pore structure decreases; as a result, the error tends to be greater.

Therefore, mercury porosimetry method has limited measure range from about 10 nm to 100 µm with this shale sample test. It can be concluded that pores diameter of this shale sample are mainly smaller than 100 nm, and pore diameter with 30 nm and below contribute more than half of total pore volume. But there is no relative correct result of pore size distribution with pore diameter less than 10 nm. That is the limitation and disadvantage of mercury porosimetry, when testing nanometer scale porous materials.

2.2.1.2 *Nitrogen Gas Adsorption*

Since the low permeability refers to the nanoscale pores, another test with higher accuracy has been implemented. As one of the methods to test pore size distribution, adsorption test is widely used in pores research, especially micropores and mesopores. Adsorption is defined as the adhesion of gas, liquid, or dissolved solid's atoms, ions, or molecules (adsorbate) to a surface of adsorbent. In laboratory test, nitrogen gas is selected as adsorbate, and shale which is pounded to small pieces or powders is adsorbent. In this shale-nitrogen system, absorbance is the function of gas pressure when temperature is fixed. This relationship between absorbance and pressure is called adsorption isotherm, which is commonly adopted in research of gas adsorption.

Many adsorption theories aim to explain adsorption isotherm of gas molecules on a solid surface. As one of the most important adsorption theories, Brunauer-Emmett-Teller (BET) (Brunauer, Emmett, & Teller, 1938) theory took multi-layer adsorption into account, and usually used adopts gases, like nitrogen, carbon dioxide, etc. as adsorbates to determine the surface area data. The reliability of BET test results is the relatively high, thus BET method is the most widely used in related industry and most of the industry standards are established based on Equation 2.3:

$$\frac{1}{V_a\left(\frac{p_o}{p}-1\right)} = \frac{c-1}{c\cdot V_m}\left(\frac{p}{p_o}\right) + \frac{1}{c\cdot V_m} \qquad\qquad Equation\ 2.\ 3$$

Where,

V_a-adsorbed gas volume;

p-equilibrium pressure;

p_o-saturation pressure of adsorbates at the temperature of adsorption;

c-BET constant;

V_m-the monolayer adsorbed gas volume.

This form of BET equation, Equation 2.3, can be plotted as a line where equilibrium related pressure $\frac{p_o}{p}$ is abscissa axis; BET function $\frac{1}{V\left(\frac{p_o}{p}-1\right)}$ is vertical axis; $\frac{c-1}{c\cdot V_m}$ is the slope; $\frac{1}{c\cdot V_m}$ is Y-intercept. If BET plot drawn in above coordinate system is a line, BET equation holds water in this case.

To analyze pore size distribution of adsorbent, Barrett-Joyner-Halenda (BJH) method is adopted in this test (Qu et al., 2010). BJH method directly calculates the pore size distribution by nitrogen gas desorption isotherm. Desorption isotherm is selected because adsorption state is more stable in the relative pressure of desorption isotherm. Like mercury porosimetry, BJH method also adopts cylindrical pore model for pore size analyzing.

High quality BET surface area and pore size distribution based on the BJH gas adsorption theory were calculated automatically while the tests were going on. The adsorption isotherm is showed on Figure 2.7. In low pressure area, adsorption curve with low adsorption volume indicates small adsorption quantity, which means this shale sample has no or few micropores (Sing et al., 1985) . In the middle section, the higher the relative pressure, the more the amount of adsorption. In high pressure area, when the relative pressure is closed to 1, the adsorption appears in macropores, adsorption curve increases sharply, and the number of adsorbed layers approaches infinite.

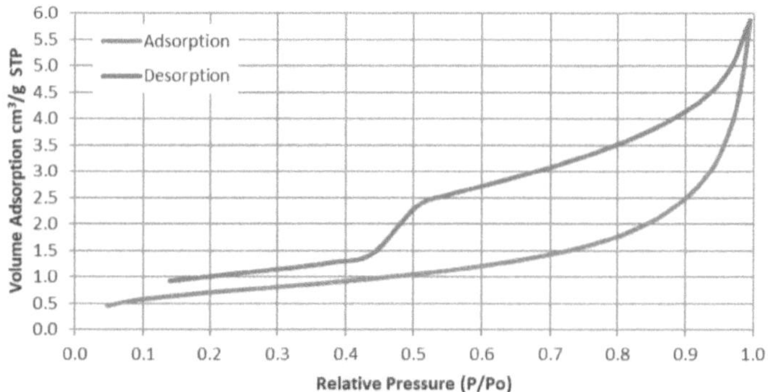

Figure 2. 7: Adsorption isotherm plot (Fang, 2017)

Desorption isotherm curve do not coincide with the adsorption isotherm curve when the relative pressure is above 0.4, which is called adsorption hysteresis. Desorption isotherm curve is located at the top of the adsorption isotherm curve. This is due to the adsorbate capillary condensation occurred in that pressure area on porous materials. This adsorption hysteresis phenomenon is related to the shape and size of the pores. According to classic theory of adsorption hysteresis hoop, the dive of desorption isotherm at relative pressure between 0.4 and 0.6 indicates that the shape of pores mainly are cylindrical and slit.

In general, the adsorption hysteresis hoop is closed. However, this adsorption isotherm curve is not closed in low pressure area. One reason may be gas escape during the adsorption test; Alternatively, it may be because of poor sample treatment before test.

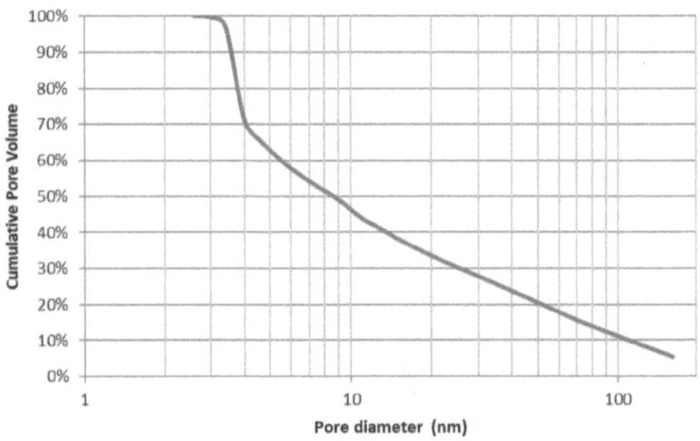

Figure 2. 8: Cumulative pore volume curve of BJH desorption pore distribution (Fang, 2017)

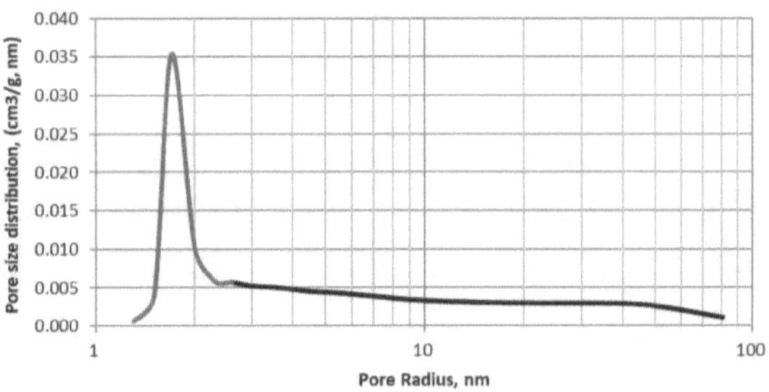

Figure 2. 9: BJH desorption pore volume distribution (Fang, 2017)

After BJH analyzing treatment, desorption cumulative pore volume and desorption pore size distribution can be obtained, Figure 2.8 and 2.9 respectively. Figure 2.8 indicates that mesopores and macropores contribute 80% and 20% to the total pore volume, respectively. It is worthwhile to note that pores with diameter less than 10nm occupy 55% of total pore volume, which cannot be measured by mercury porosimetry.

Figure 2.9 pointed that the pore radius ranges from 1.3 nm to 80.8 nm. However, the BJH method is not applicable for micropores, so the peak pore size distribution in small pore radius area (red part of curve) is incorrect. Moreover, when relative pressure is close to 1, the requirement of gas pressure accuracy is so high that the large pore radius at the end of pore size distribution curve is not accurate, neither.

After both ends revision, the pore size distribution is shown on Figure 2.10. The curve indicates that pore size mainly distributes in mesopores range. Small mesopores with diameter less than 10nm have a higher fraction. The larger the pore size, the lower the amount. The number of pores with diameters larger than 50nm decreases slowly and then faster.

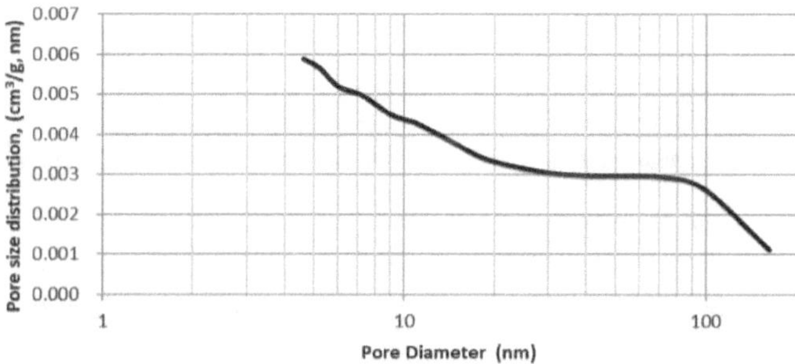

Figure 2. 10: Pore size distribution after both ends revision (Fang, 2017)

2.2.1.3 Comparison and Discussion

Mercury porosimetry and nitrogen gas adsorption are both widely used for the determination of pore size distribution. As shown in the Table 2.1, both methods adopt cylindrical pore model. However, the testing duration, effective measuring range, obtained total pore volume and average pore diameter are diverse from each other.

Table 2. 1: Comparison of mercury porosimetry and nitrogen gas adsorption (Fang, 2017)

	Mercury Porosimetry	Nitrogen Gas Adsorption
Elapsed time	Less than 1 hour	11 hours
Effective range of pore diameter	10 nm - 100 μm	5 nm - 150 nm
Pore Volume (mm^3/g)	5.36 (7.33 nm - 100 μm)	9.206 (1.7 nm - 300 nm)
Average pore diameter (nm)	20.01	9.33

Mercury porosimetry is not accurate for measuring the nanometer scale pores, because higher pressure will cause great error data due to rock structure damages and particles breaking. This method is more suited for macropores measurements. In contrast, nitrogen gas adsorption is more accurate to measure mesopores. The comparison of total pore volume indicates that pores of diameter less than 7.33 nm contribute 3.84 mm^3/g pore volume, which equals about 70% total volume of pores with diameter larger than 7.33 nm. Even if the measurement error is neglected, small pores which mercury porosimetry cannot measure play a decisive role in total pore volume.

Comparing pore size distribution of two methods in Figure 2.11, the distribution profiles trend of desorption and intrusion is the same below 70 nm. It is because that the adsorption and extrusion mechanism is more controlled by the relative wider portions of the pores, while desorption and intrusion processes are essentially a pore throat controlled (Dullien & Dhawan, 1974; Mason, 1982).

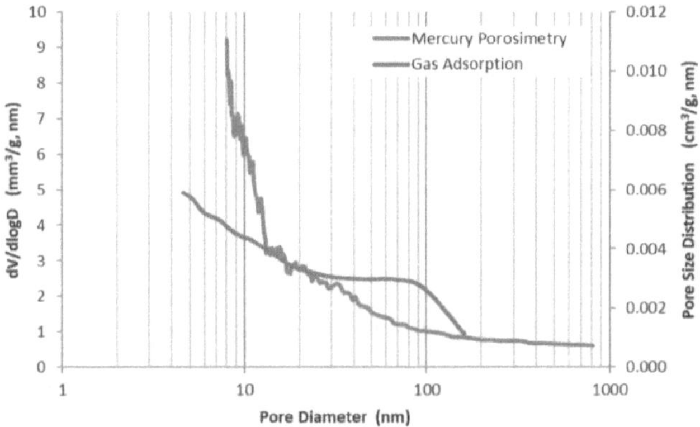

Figure 2. 11: Pore size distribution of two methods (Fang, 2017)

However, the difference in profiles shows the varying extent of pore and pore throat controlled. Here, highlighting the measurement range of two methods that gas adsorption can show smaller pore size data, while mercury porosimetry can reach micron scale pores.

Therefore, though mercury porosimetry is commonly used for conventional reservoir rocks, nitrogen gas adsorption is better suited for shale with nanometer scale pores to analyze pore properties (total pore volume, pore size distribution, etc.), especially for lower porosity lacustrine shale.

2.2.2   Scanning Electron Microscope

Scanning Electron Microscope (SEM) is a type of electron microscope that can scan sample with a focused electron beam to generate images of the sample's surface topography and analyze the chemical composition of selected points on sample's surface if coupled with Energy-dispersive X-ray Spectroscopy (EDS).

SEM is one of most commonly used methods for micro structure investigation. It can provide directly perceived images of samples with micron, nanometer scale. Hence, SEM is selected to observe and analyze shale nanoscale pore structure. SEM images of shale pores structure will help us with a better understanding of shale at fundamental level (Wu & Aguilera, 2012).

A lot of efforts have been spent on shale SEM images analysis by many scholars. The universality and persuasiveness of some images, however, remain to be discussed. The selection of sample surface for observation is of great importance to investigate pore structure. For example, some visible concave pores and cracks on an arbitrary fractured surface may be produced by artificial broken. They do not really exist in the reservoir when spaces are filled by opposing surface bulges under the strata pressure.

Unlike conventional gas bearing rock, shale lithology has great directivity. The same shale core samples have been tested with nitrogen medium to measure its permeability in lab. It is a remarkable fact that there is no result of permeability in vertical direction. In other words, the vertical permeability can be regarded as null. In horizontal direction, results of permeability are extremely low that equivalent to $k = 0.172 \times 10^{-3}\ mD$. Pore properties of shale are different in horizontal and vertical direction. Horizontal fractured surface and vertical cross section of shale sample have been observed in SEM.

2.2.2.1　SEM Images of Horizontal Surface

The surface for testing is a fractured horizontal surface. After preparing the small sample, shale's surface topography can be observed. Figure 2.12 indicates no obvious fractures and pores in this surface. However, the outline of minerals with varying dimensions can be recognized. Suspect fractures and pore space can be found along with minerals external profile marked by red circles and inside some minerals marked by the white circles.

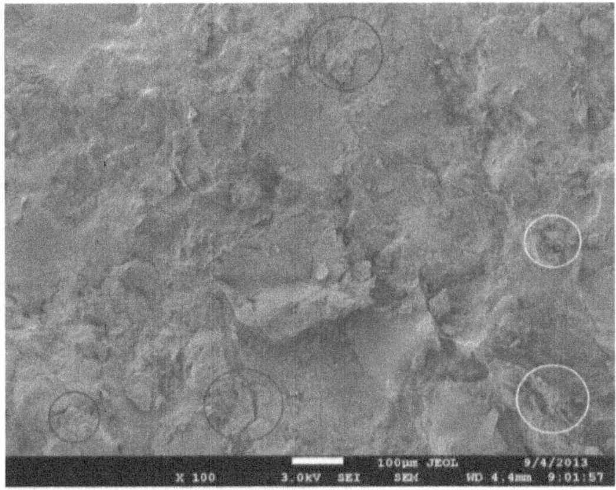

Figure 2. 12: SEM images of shale horizontal surface (Fang, 2017)

With the increasing magnifications, the content in Figure 2.13 becomes clear. Several pyrite framboid which prevails throughout shale all over the world are distributed around the black organic matter. Two micro cracks are visible at the border between two minerals, of which one micro crack spreads across the organic matter.

In Figure 2.13, six types of pore space are grouped into two general categories on the basis of size: nanoscale pores (the pore size is less than one micron) and micron scale pores (the pore size is larger than one micron). Nanoscale pores and cracks exist both within grains of mineral and between them. Pyrite framboid are so widely distributed in shales that formed two types nanoscale pores. The pores located along organic matter boundary and inside flaky minerals are generally in nanoscale. However, the crack at organic and minerals boundary can develop to a width larger than one micron. All six pore types are showed in Table 2.2. The numbered items in the table are marked in Figure 2.13 with the same numbers.

Figure 2. 13: SEM images of shale horizontal surface with pyrite framboid, organic matter and micro fractures (Fang, 2017)

Table 2. 2: Form, distribution and size of pores marked in Figure 2.13 (Fang, 2017)

Category	Form	Location	Distribution	Item
Nanoscale	Pore	Pyrite framboid	Between pyrite particles and rock matrix	1
	Pore	Pyrite framboid	Pyrite intergranular	2
	Crack	Organic matter	Between organic matter and rock matrix	4
	Pore	Minerals	Inside pieces of same mineral	6
Micron scale	Crack	Organic matter	Across organic matter	3
	Crack	Minerals	Between different minerals	5

Except for item 3 and 4 of pore spaces in organic matter, nanopores inside organic matter is one of the most widely exist pore types in the shale, which is considered in dual porosity research of shale. However, not all of the organic matters develop nanopores inside. It depends on the maturity of organic matter. The organic matter with low maturity has few pores. Figure 2.14 shows that the organic matter in lacustrine shale, Yanchang shale have no "bigger" nanopores (diameter larger than 50 nm), while marine shale, taking Barnett shale for example, intraparticle organic nanopores are well developed.

These nanoscale pores constitute the most widespread and numerous pore type in the Barnett Shale (Loucks, Reed, Ruppel, & Jarvie, 2009). Organic matter in lacustrine shale may show a lower maturity than that of marine shale.

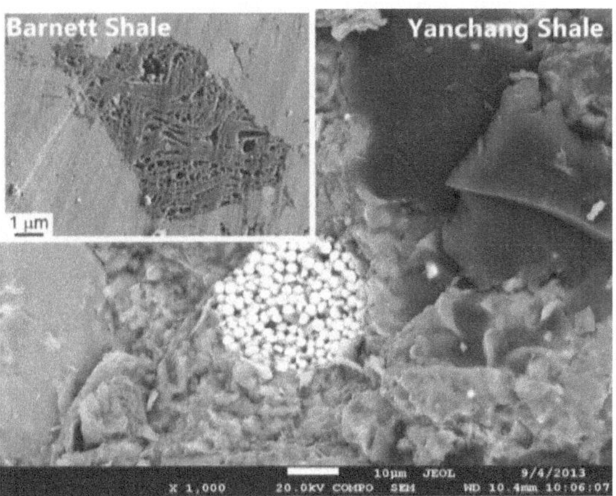

Figure 2. 14: Comparison of organic matter in Barnett shale and Yanchang shale (Loucks et al., 2009)

In summary, the main pore types of this shale sample are micro fractures in organic matter and in the border between two minerals. The length of micro fractures could reach 100 µm, and width ranges from 0.1 µm to 4 µm. However, micro cracks have not connected to each other and neither did scattered pores.

2.2.2.2   SEM Images of Vertical Cross Section

As shown in the Figure 2.15, microscale cracks with 20 µm to 500 µm long are well developed. Meanwhile, pores with less than 10 µm diameter are found, which are likely produced by the mineral dissolution. Minerals are pressed into sheets, which are shown as horizontal bar cross sections. Sporadic distributed white spots are pyrite framboid with different scales.

Figure 2. 15: SEM image of shale vertical cross section (Fang, 2017)

Unlike the cracks found in the horizontal surface, cracks connected to each other composing longer crack as shown in Figure 2.16. Microscale cracks and pressed minerals are of the same direction, both tending towards horizontal. Pieces minerals and cemented fillers with more nanopores are easier to develop and connect micro cracks.

Figure 2.17 can be divided into three parts according to the shape of cracks and minerals. Minerals in upper area are mostly blocks, while minerals in lower area are flat, slim, dense and compacted. The middle area develops long fractures. This difference can be seen without microscope by the layers in shale cores.

This image indicates the change of sedimentary conditions that lower part is formed under higher pressure. Thus, lower minerals become flatter, smaller and more fragmentary. Meanwhile, micro cracks and pores are well developed in such pieces of minerals. Because of the change in sedimentary conditions, the minerals in transition area grow in different states, which results in the easy formation of longer fractures in horizontal direction in this area.

Figure 2. 16: Magnified SEM image of vertical cross section (Fang, 2017)

Figure 2. 17: SEM image of vertical cross section indicating sedimentary situation (Fang, 2017)

The fragmentation of minerals creates more pores and micro fractures, as shown in Figure 2.18. It indicates that the porosity of shale is largely dependent on the appearance of minerals. In addition, the nanoscale pores in small pieces minerals and cemented fillers increase significantly the surface area for gas adsorption.

Figure 2. 18: SEM image of vertical cross section (Fang, 2017)

Figure 2. 19: SEM image for the comparison of horizontal surface and vertical cross section. Left side is horizontal surface, while right side is vertical cross section with obvious fractures and layers (Fang, 2017)

In summary, SEM images of vertical cross section show more micro fractures and pores than that of horizontal surface in Figure 2.19. Combining two direction images research, we can conclude that micro cracks are developed in shape of flat, paralleling to the horizontal plane, with up to 500 µm long. Macropores and mesopores mainly exist in intergranular spaces of minerals and in the border between two minerals. Small pieces minerals are easier to develop and connect micro cracks.

The connected micro fractures in horizontal direction increase the permeability of that direction. Shales rarely have cracks in vertical direction. Thus, the vertical permeability of shale generally is null. Micro fractures can effectively improve the seepage capacity of shale reservoir, after hydraulic fracturing treatment, they could become the main seepage channel in micro scale. Therefore, extensive developed micro fractures in shale are conducive to the storage of gas, and can significantly improve the permeability of reservoir as well.

Table 2. 3: Mineral content of weight and area (Fang, 2017)

Mineral	Mass fraction (%)	Area fraction (%)	Area (micron)	Grain Count
Quartz	20.765	21.300	27226177.28	31
K-Feldspar	12.216	12.920	16515112.36	9
Albite	24.916	25.850	33042697.33	8
Other Silicates	1.312	1.049	1340292.82	2
Carbonate-Sulfate	1.125	1.128	1441943.06	1
Clay-Illite-type	3.622	3.516	4494249.99	6
Clay-Kaolinite-type	16.612	17.367	22199123.39	28
Clay-Fe-Mg-type	13.949	12.639	16155567.13	8
Clay-Phosphate-Mix	1.990	1.691	2161178.89	2
Apatite-Phosphates	0.680	0.577	737959.27	6
Phosphate-REE-Min	0.003	0.002	1945.03	1
Pyrite-Sulfides	1.774	0.965	1232868.56	25
Fe-Mn-Oxides	0.101	0.052	66389.70	2
Ti-Fe-Oxides	0.278	0.158	201338.00	1
Zircon	0.023	0.013	16711.54	2
Invalid	0.636	0.775	990780.91	4
Total	100.000	100.000	127824335.27	118

2.2.3 Mineral Content and Composition

The statistics related to mineral content was presented by X-Ray map. Table 2.3 indicates all the mineral mass fraction and area fraction. It is clear that albite content is the highest which is 24.9%, followed by quartz content with a value of 20.7%. The total of feldspar, K-feldspar and albite is up to 37.1%, being the most mineral in this lacustrine shale sample. It can be obviously seen in Figure 2.20 that Feldspar fraction is higher than quartz. Meanwhile, Clay fraction is not low neither. Higher brittle minerals, including quartz, K-feldspar, Albite and other silicates, with 59% content and clay occupying 36% are the two main constituents of shale.

X-Ray map can also provide the information of minerals distribution. Figure 2.21 showed that quartz distributed dispersedly and widely; clay came out as big block and had some extent of connectivity.

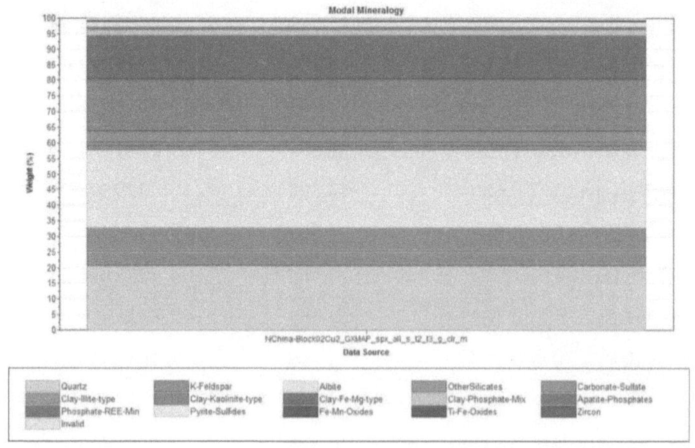

Figure 2. 20: Mineral weight content (Fang, 2017)

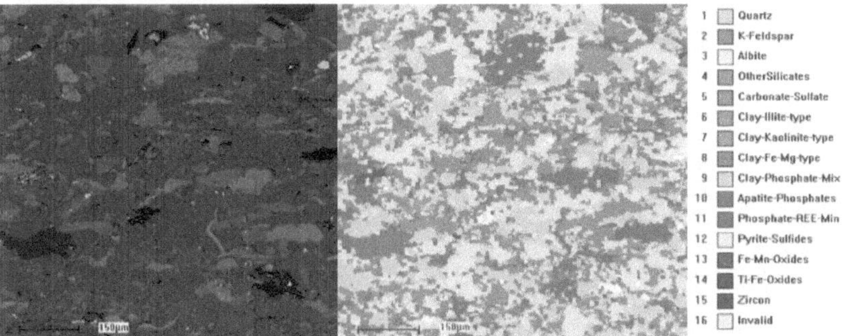

Figure 2. 21: Mineral distribution determined by Grid X-ray map response BSE image (Fang, 2017)

2.3   Summary

The results from laboratory described the characterization of samples. Among them, the pore distribution was investigated through two methods, mercury porosimetry and nitrogen adsorption, which have completely different precision and measuring range. In correlation with the SEM, which could reveal the micro-structure of matrix, most pores in shale sample were nano-magnitude, making difficulty in exploiting shale gas. However, the SEM also pointed out that fracturing in horizontal direction is much more than in vertical. It implied that the treatment of fracture in horizontal direction is the better way to increase permeability.

Furthermore, the X-Ray analyzed the composition and contend of mineral. Fraction of the feldspar is higher than quartz. Generally, the greater the content of quartz, the easier to be fractured. Nevertheless, the reconsolidation also happened more frequently after having been fractured, which will be demonstrated in Chapter 3. Another interesting perspective is that total content of feldspar is up to 37.1%, as interpreted, which has a positive effect on desorption. Further details and discussion will be developed in Chapter 4.

3   Reconsolidation

Researches have listed numbers of factors and reasons which can cause the low yielding of shale gas. Mostly, it was attributed in low permeability. Another spoken frequently is the reconsolidation of silica in the shale resource matrix. Although the phenomenon of reconsolidation has been observed and researched for years in exploration of conventional formation, there is still no effective solution for the sedimentary formation with low permeability yet, more specifically shale reservoir.

The reclosed fracture observed after the processing of shale formation for more production, could be a solid proof for the occurrence of reconsolidation. After the tertiary treating, conducted for increasing the yielding, the productivity of shale was not enhanced obviously. Instead, reduction in the productivity have been detected in some cases in field. Scholars debated on whether the reclosure of fracture can be attributed to confining pressure, except for the reconsolidation of silica. Furthermore, the temperature should not be neglected at any time.

The second evidence was the disappeared proppants in the practice, even in the laboratory. After conducting hydraulic fracturing experiments, Martin Müller could not find the added proppants in the rock sample (Müller, 2017). One possible explanation might be the precipitation of dissolved silica, which covered the added proppants while the silica dissolution began gelling in the induced fractures.

The low productivity reflects the importance and necessity to study the mechanism of reconsolidation in shale for the further exploration. Every procedure of the process should be analyzed. And the factors which played a leading role have to be determined.

Theoretically and strictly, the following investigation should be implemented by using the same shale core as so for the analysis of characterization in last section. However, the ultra-low permeability limited the intent. As demonstrated in Chapter 2, the shale matrix is composed of silica and clay. Consisting of components similar with shale, diatomite cores with relative great permeability had been used to analyze the influence factors which resulted in the precipitation and gelling of dissolved silica. Because of the relative great permeability, the procedure with diatomite was more efficient than shale core in laboratory. Although the results obtained from diatomite were slightly unconvincing, the focus of this part is on the determination of influence factors and the chemical causes. The investigation in this chapter begins with the dissolution of silica and ends with reconsolidation.

3.1 Silica Dissolution

As Bhat described (Bhat & Kovscek, 1999), pH, salinity, and temperature have a great effect on the dissolution in reservoirs. Therefore, three series of experiments were presented and analyzed in this section for further details.

3.1.1 Experiment Series I: the Influence of pH

In this dissolving experiments, four brine solutions with various pH were prepared as injection fluids at room temperature. And the compositions of fluids and the corresponding initial conditions are shown in Table 3.1. Permeability, pH value and silica concentration of effluent were measured every hour.

Obtained using the four solutions described in Table 3.1, the relationship between pH of effluent and injection volume has presented in Figure 3.1. From these tests, a similar tendency could be summarized for all tests that pH decreases with the injection volume, regardless of the initial pH of the injected fluid. A reduced pH in effluent means that the hydroxide (OH⁻) in the injected solution has reacted inside the sample. Furthermore, the values became constant after reaching about 4. It is also possible that the same equilibrium pH for silica dissolution reaction was reached under these experimental conditions.

Table 3. 1: Composition of brines for Experient Series I with varying pH (Peng, 2009; Peng & Kovscek, 2011)

		1	2	3	4
Temperature	°C	20	20	20	20
pH		4	6	10	12
Na^+	ppm	3961	3961	3961	3961
Ca^{2+}	ppm	56	56	56	56
Mg^{2+}	ppm	3	3	3	3
Cl^-	ppm	6222	6222	6222	6222
Salinity	ppm	10242	10242	10242	10242

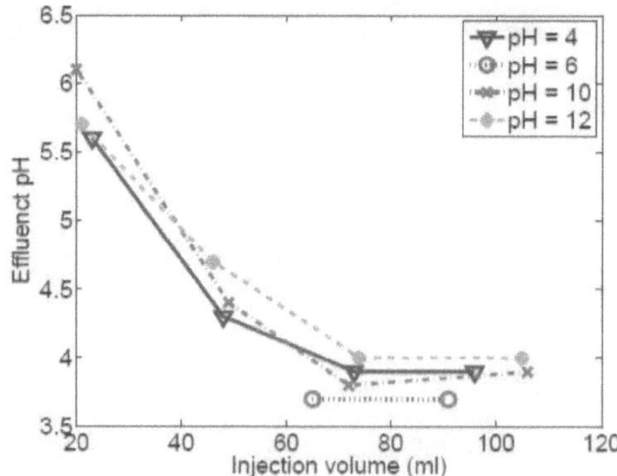

Figure 3. 1: Relationship between pH of effluent and injection volume for Experiment Series I (Peng, 2009; Peng & Kovscek, 2011)

The silica concentration of effluent was measured after the dissolution experiments. According to Figure 3.2, with the same injection volume, the greater pH of the injection fluid, the greater silica concentration. Consequently, it can be deduced that alkaline condition (pH > 7) is good for silica dissolution.

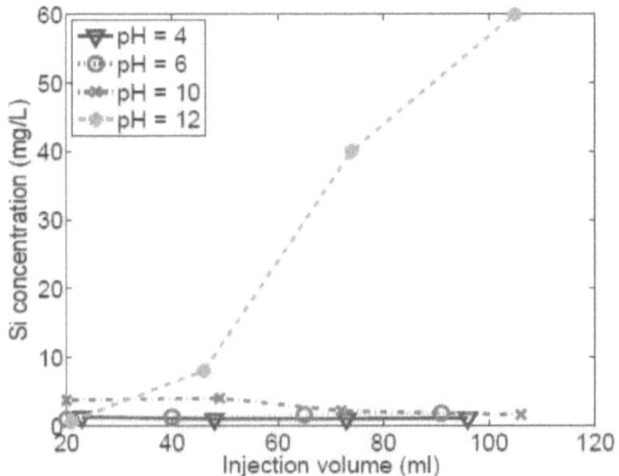

Figure 3. 2: Relationship between silica concentration of effluent and injection volume for Experiment Series I (Peng, 2009; Peng & Kovscek, 2011)

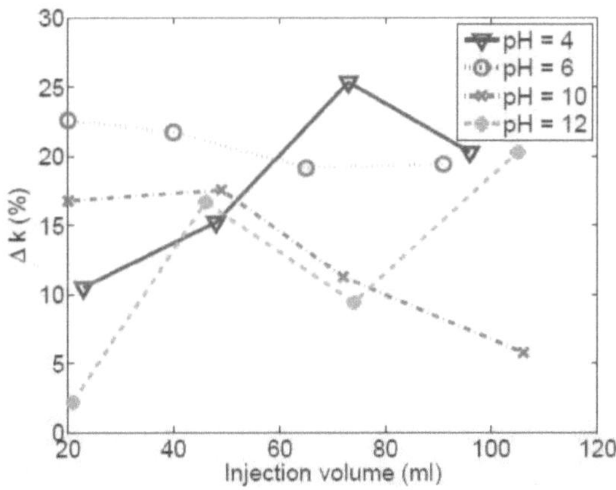

Figure 3. 3: Relationship between the change in permeability and injection volume for Experiment Series I (Peng, 2009; Peng & Kovscek, 2011)

As shown in Figure 3.3, an increase in permeability was observed after the injection. Moreover, the values of change in permeability varied and the change did not follow a constant tendency for the change.

As explained by Diabira, dissolution in diatomite matrix is a slow process and the formation of moderately homogeneous rock cannot be changed substantially unless a large volume of fluid is injected (Diabira, Castanier, & Kovscek, 2001). The injection volume was less than 120 ml and injection time was about 4 hours, which explained why the permeability did not change much during the conducting. Another possible reason for the irregular change was related to the preparation of sample by flushing. Some fine grains maybe be brought out of the matrix through low salinity solution (Peng, 2009; Peng & Kovscek, 2011).

3.1.2   Experiment Series II: the Influence of Salinity

To study the effects of salinity on silica dissolution, experiments were conducted using three brine solutions with different salinity under the conditions of room temperature (20 °C) and constant pH (pH = 10). The related information of the solutions is summarized in Table 3.2.

Table 3. 2: Compositions and conditions of brines for Experient Series II with varying salinity (Peng, 2009; Peng & Kovscek, 2011)

		1	2	3
Temperature	°C	20	20	20
pH		10	10	10
Na^+	ppm	131	1900	7798
Ca^{2+}	ppm	56	56	56
Mg^{2+}	ppm	3	3	3
Cl^-	ppm	310	3041	12144
Salinity	ppm	500	5000	20001

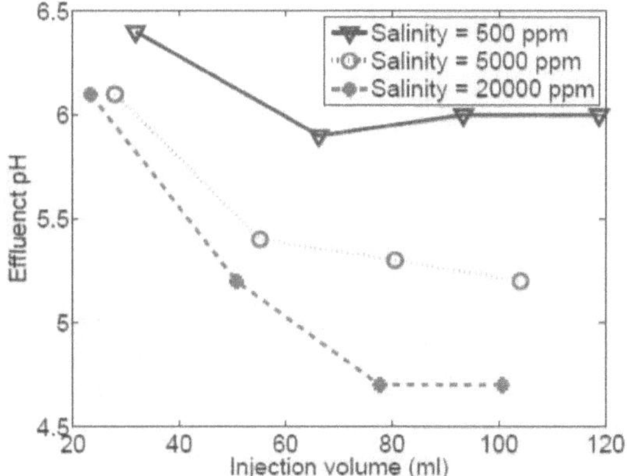

Figure 3. 4: Relationship between pH of effluent and injection volume for Experiment Series II (Peng, 2009; Peng & Kovscek, 2011)

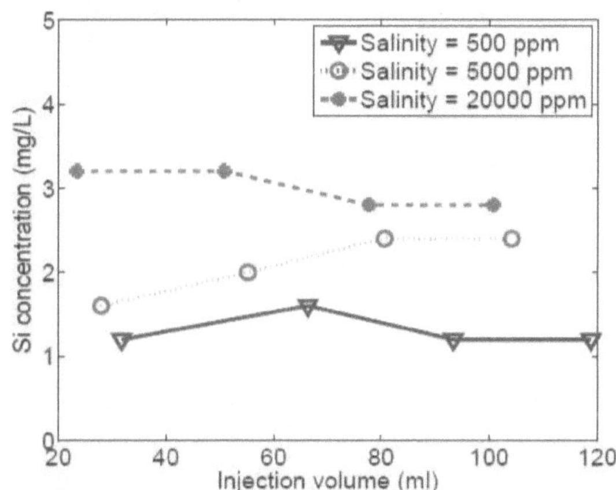

Figure 3. 5: Relationship between silica concentration of effluent and injection volume for Experiment Series II (Peng, 2009; Peng & Kovscek, 2011)

At room temperature under a constant pH condition, low salinity solution resulted in

greater effluent pH, as shown in Figure 3.4. The reason may be that the dissolution was light in the solution with low salinity, so amounts of hydroxide (OH⁻) was not consumed readily and then the pH value remained high.

In agreement with the conclusion above, silica concentration in effluent pointed that the brine solution with higher salinity caused stronger dissolution, as demonstrated in Figure 3.5.

As shown in Figure 3.6, permeability did not change much with the injection volume, which was similar to the results of the pH test. As interpreted in Experiment Series I, silica dissolution was a slow process and the rock matrix cannot be greatly changed unless a large volume of fluid was injected (Diabira et al., 2001).

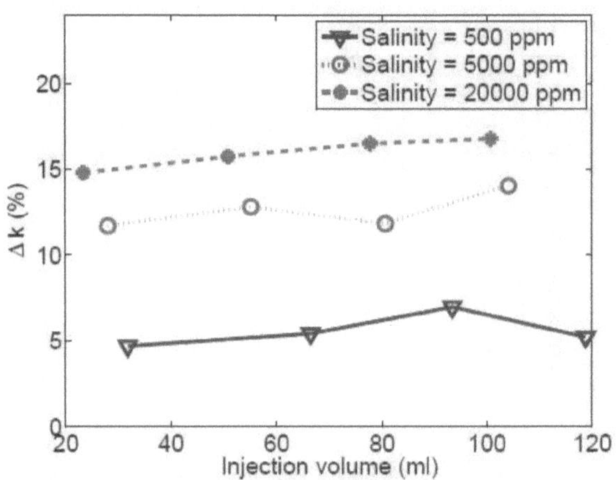

Figure 3. 6: Relationship between the change in permeability and injection volume (Peng, 2009; Peng & Kovscek, 2011)

3.1.3   Experiment Series III: the Influence of Temperature

Under the experimental conditions described in Table 3.3, the influence of temperature on dissolution was investigated in this section. With different degrees, pH and silica concentration of effluent as well as change of permeability were listed as well.

Based on Figure 3.7 and 3.8, it is clear that pH reduced much more slowly at elevated temperature, while the silica concentration became much greater. It also suggested that dissolution was promoted strongly at elevated temperature.

Table 3. 3: Composition of brines for Experiment Series III with varying temperature (Peng, 2009; Peng & Kovscek, 2011)

		1	2	3	4
Temperature	°C	20	120	180	230
pH		10	10	10	10
Na^+	ppm	3961	3961	3961	3961
Ca^{2+}	ppm	56	56	56	56
Mg^{2+}	ppm	3	3	3	3
Cl^-	ppm	6222	6222	6222	6222
Salinity	ppm	10242	10242	10242	10242

The explanation of slight changing in permeability at high temperature, as indicated in Figure 3.9, was presented by Diabira (Diabira et al., 2001), which has stated above. In Comparison with results of experiments I and II involving varying pH and salinity, temperature plays a more significant role on the silica dissolution process.

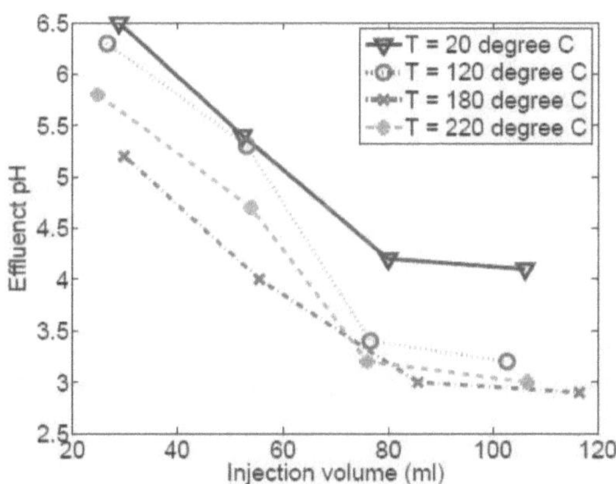

Figure 3. 7: Relationship between pH of effluent and injection volume for Experiment Series III (Peng, 2009; Peng & Kovscek, 2011)

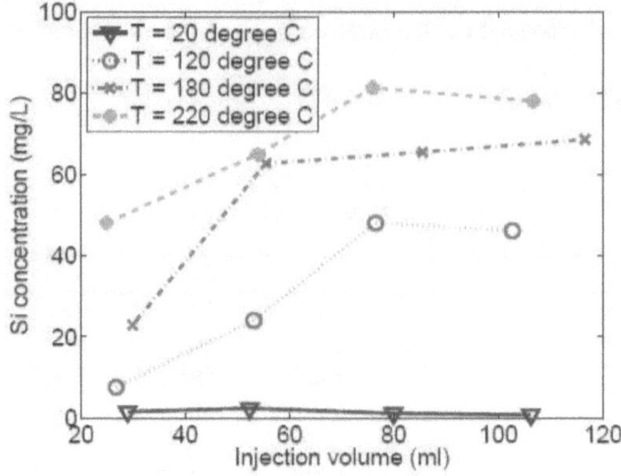

Figure 3. 8: Relationship between silica concentration of effluent and injection volume for Experiment Series III (Peng, 2009; Peng & Kovscek, 2011)

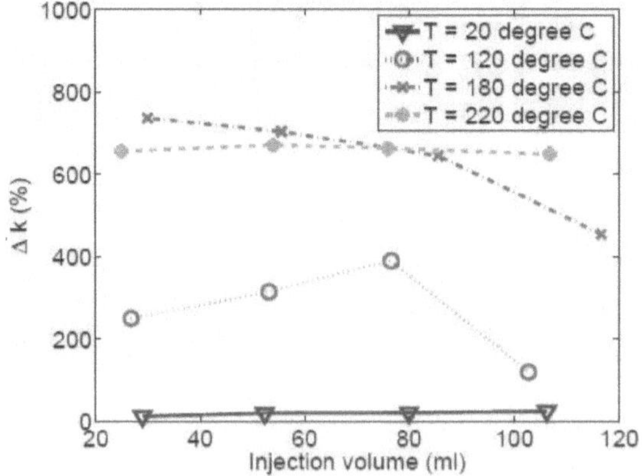

Figure 3. 9: Relationship between the change in permeability and injection volume for Experiment Series III (Peng, 2009; Peng & Kovscek, 2011)

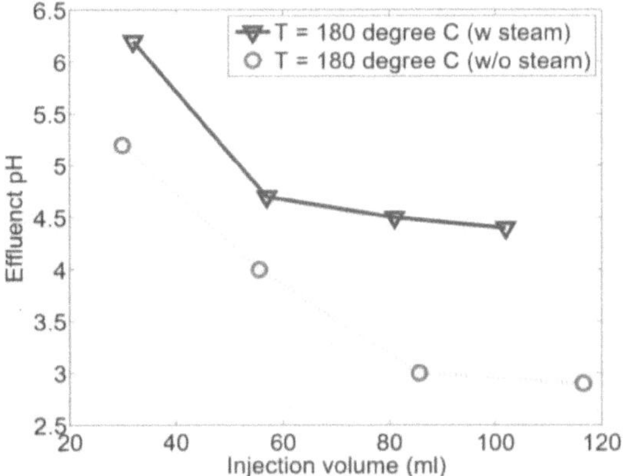

Figure 3. 10: Relationship between pH of effluent and injection volume at 180 °C (Peng, 2009; Peng & Kovscek, 2011)

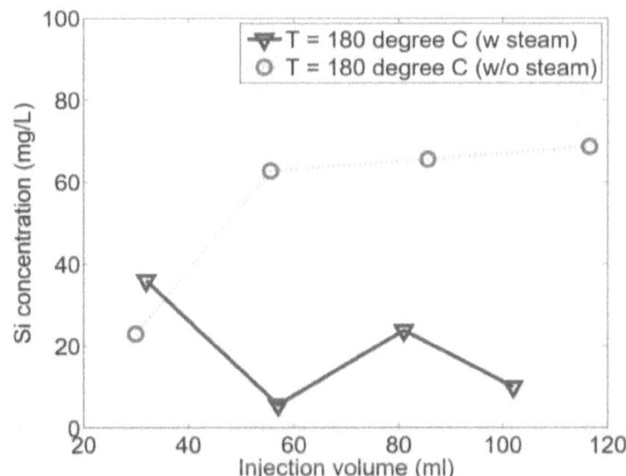

Figure 3. 11: Relationship between silica concentration of effluent and injection volume at 180 °C (Peng, 2009; Peng & Kovscek, 2011)

At temperatures higher than 100 °C, temperature might affect the dissolution process by

the presence of steam. An additional test was conducted at 180 °C to investigate the

influence of steam. As shown in Figure 3.10 and 3.11, the extent of dissolution without steam is much greater than that with steam. Therefore, steam did not promote the process of silica dissolution.

3.2   Reconsolidation Experiments

From the preceding Experiment Series I – III, the optimal parameters were drawn and applied in this section of reconsolidation investigation. Because of the great influence of temperature and pH on dissolution, a pH value of 10 and 200 °C were assigned to the experimental condition of effluent pH and temperature, respectively. Under these experimental conditions, tests were conducted to clarify the mechanism of reconsolidation. More details were presented in Table 3.4.

Table 3. 4: Experimental conditions for Experiment Series IV (Peng, 2009; Peng & Kovscek, 2011)

Test	Fractures		Injection	PH	Temperature	Confining pressure
	Quantity	Direction				
1	1	direction of flow	basic brine	10	200 °C	400 psi
2	1	direction of flow	buffer	10	200 °C	400 psi
3	2	direction of flow	buffer	10	200 °C	400 psi

3.2.1   Experiment Series IV: the Mechanism of Reconsolidation

As shown in Figure 3.12, the sample in Reconsolidation Test 1 consisted of two halves. The fracture oriented in the direction of injection. Similar to dissolution experiments, the pH and silica concentration of the effluent were monitored during the test, and the same basic brine with pH value of 10 was injected in Test 1. During the injection with a duration of 100 hours, the core was scanned by CT.

From Figure 3.13, pH of effluent sank sharply to 2 at the beginning of injection. Nevertheless, it remained below 3 for only a short period. Then it gradually enhanced to a plateau. As shown in Figure 3.14, the silica concentration declined also greatly during the sharp decreasing of pH of effluent. It indicated that injected fluid dissolved silica strongly at the first. After that, concentration varied from 150 mg/L to 300 mg/L. Correlating with pH and concentration, permeability exhibited the same tendency, as illustrated in Figure 3.15. The CT images for Test 1 shown in Figure 3.16 demonstrate the same tendency as well. The reduction at the beginning could be attributed to the debris between two halves, which is more easily to be dissolved in alkaline fluid.

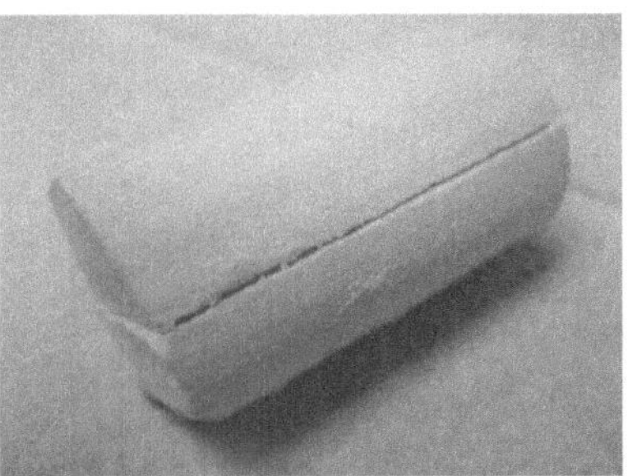

Figure 3. 12: Sample consisting of two haves for reconsolidation Tests 1 and 2 (Peng, 2009; Peng & Kovscek, 2011)

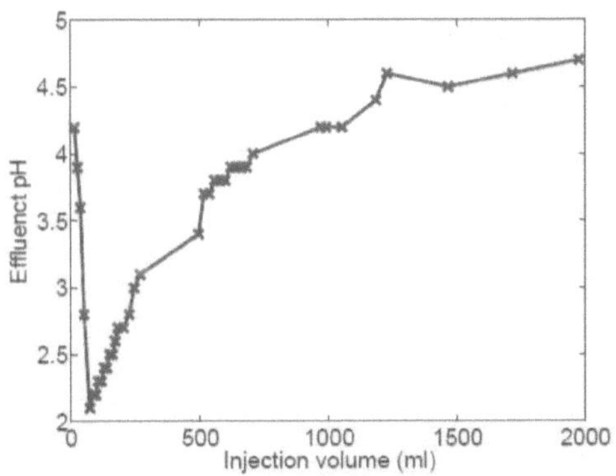

Figure 3. 13: Relationship between pH of effluent and injection volume for Reconsolidation Test 1 (Peng, 2009; Peng & Kovscek, 2011)

The Reconsolidation Test 2 was performed under the same conditions as Test 1, except that the injection fluid was replaced by buffer solution composed of $NaHCO_3$ and

Na$_2$CO$_3$. Small amounts of acid or base were added in the solution to keep the pH of this buffer solution constant (Peng & Kovscek, 2011).

Unlike the behavior in Test 1, pH was mostly greater than 7 in Test 2, as shown in Figure 3.17. It increased from the beginning stably. After the value reached about 10, it remained almost constant. Figure 3.18 showed the correlation between the injection volume and the silica concentration which was relatively high when pH was relatively low. As shown in Figure 3.19, the permeability enhanced constantly while the fluid was injected (Hoefner & Fogler, 1988). A possible reason was that the injection fluid was always alkaline.

The porosity distribution in the sample before and after Test 2 was compared in Figure 3.20. The sample in Test 1 was affected by base brine and the porosity increased consequently. Theoretically, the porosity should rise in Test 2. However, it decreased in reality. It maybe be attributed to the reconsolidation and confining pressure.

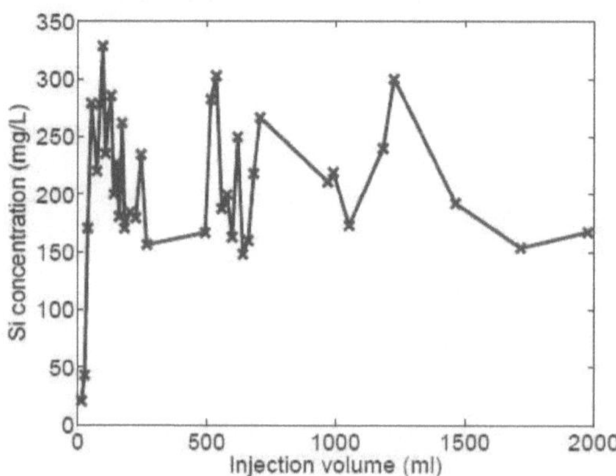

Figure 3. 14: Relationship between silica concentration of effluent and injection volume for Reconsolidation Test 1 (Peng, 2009; Peng & Kovscek, 2011)

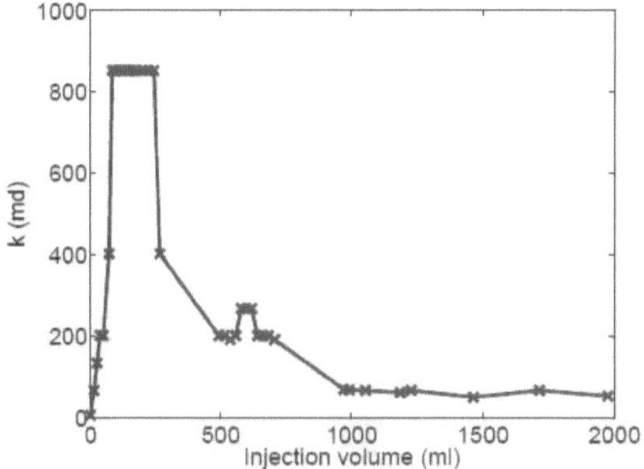

Figure 3. 15: Relationship between permeability and injection volume for Reconsolidation Test 1 (Peng, 2009; Peng & Kovscek, 2011)

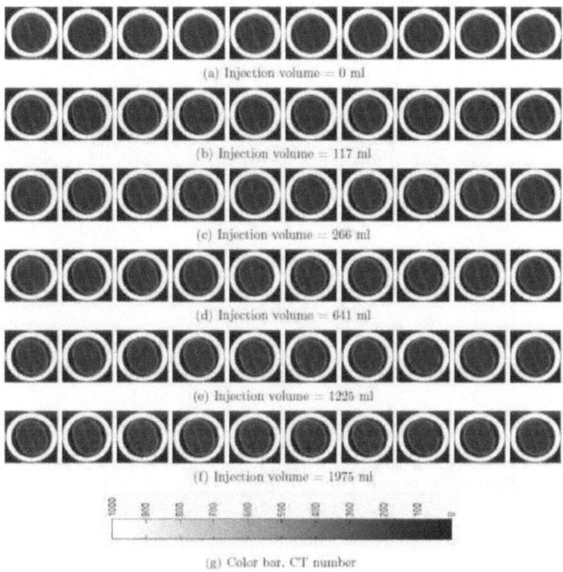

Figure 3. 16: CT images for Reconsolidation Test 1 (Peng, 2009; Peng & Kovscek, 2011)

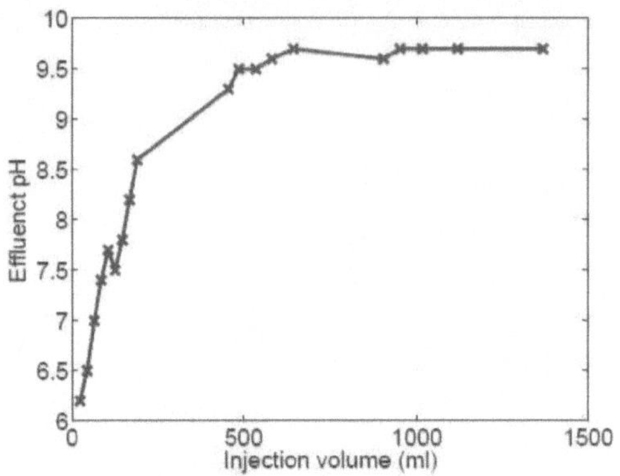

Figure 3. 17: Relationship between pH of effluent and injection volume in Reconsolidation Test 2 (Peng, 2009; Peng & Kovscek, 2011)

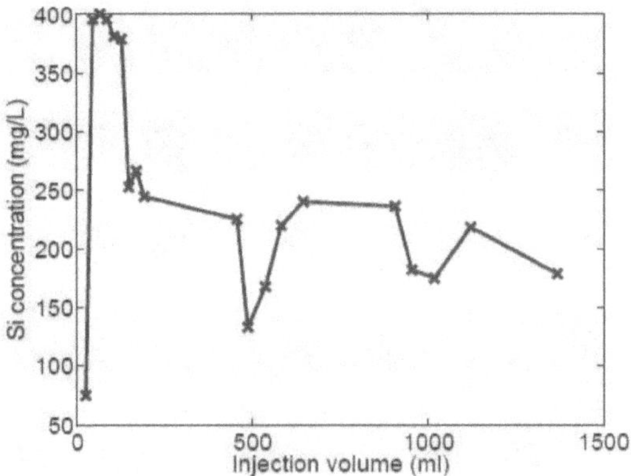

Figure 3. 18: Relationship between silica concentration of effluent and injection volume in Reconsolidation Test 2 (Peng, 2009; Peng & Kovscek, 2011)

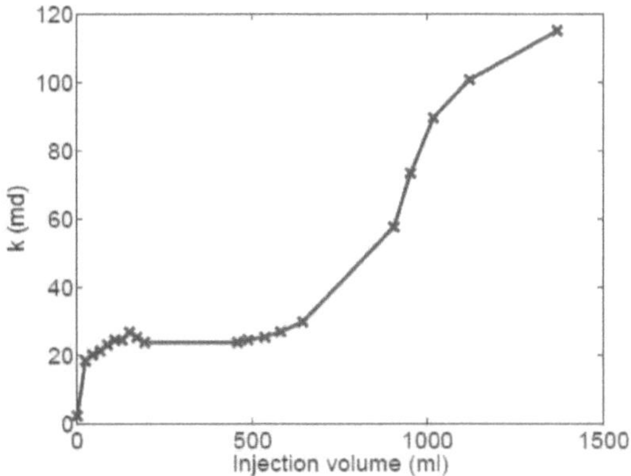

Figure 3. 19: Relationship between permeability and injection volume in Reconsolidation Test 2 (Peng, 2009; Peng & Kovscek, 2011)

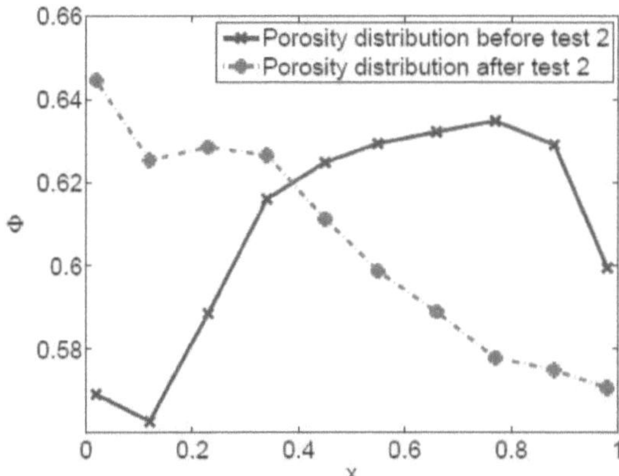

Figure 3. 20: Porosity of distribution in the sample in Reconsolidation Test 2 (Peng, 2009; Peng & Kovscek, 2011)

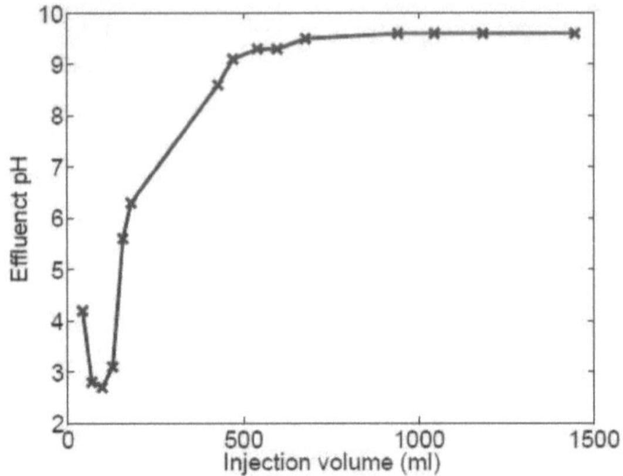

Figure 3. 21: Relationship between pH of effluent and injection volume in Reconsolidation Test 3 (Peng, 2009; Peng & Kovscek, 2011)

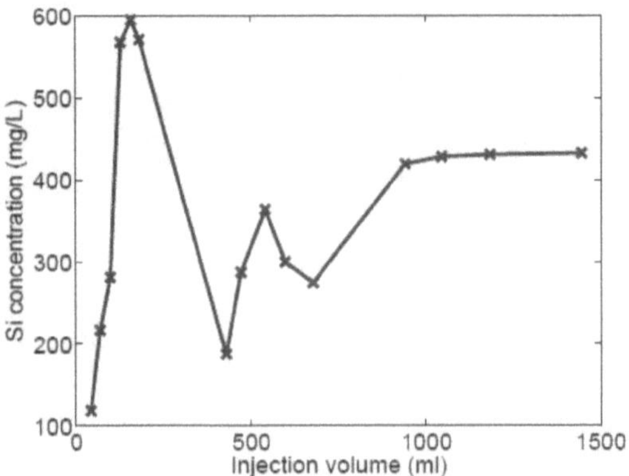

Figure 3. 22: Relationship between silica concentration of effluent and injection volume in Reconsolidation Test 3 (Peng, 2009; Peng & Kovscek, 2011)

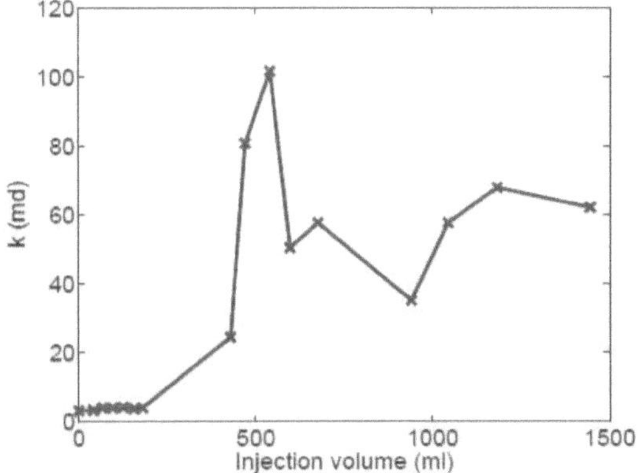

Figure 3. 23: Relationship between permeability and injection volume in Reconsolidation Test 3 (Peng, 2009; Peng & Kovscek, 2011)

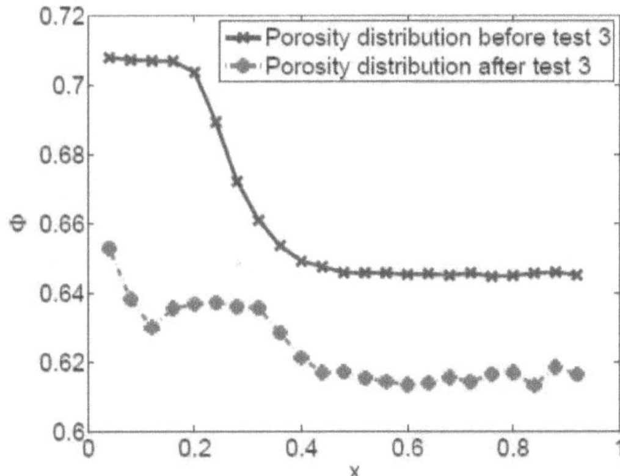

Figure 3. 24: Porosity distribution along the sample in Reconsolidation Test 3 (Peng, 2009; Peng & Kovscek, 2011)

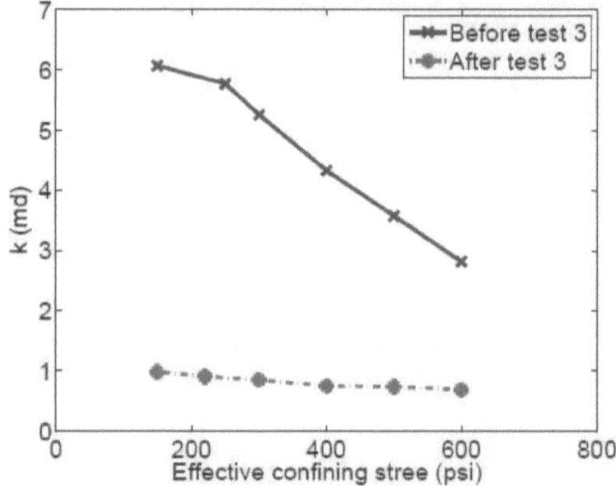

Figure 3. 25: Relationship between permeability and effective confining stress before and after Test 3 (Peng, 2009; Peng & Kovscek, 2011)

To justify the results of Test 2, Test 3 was conducted under the same experimental

conditions. In Test 3, the new core had two fractures in the lengthwise direction. Similar

to Test 2, pH increased to almost 10 subsequent to a decrease at the beginning, as

demonstrated in Figure 3.21. Silica concentration exhibited a similar behavior as pH.

When the pH reached a plateau, the extent of dissolution became stable as illustrated in

Figure 3.22. In contrast, the permeability had an obvious peak. In the second half of

injection, the permeability sank, which suggested the fractures were healed and the rock

became more consolidated. The permeability and porosity distribution in Figure 3.23 and

Figure 3.24 also proved the occurrence of reconsolidation. The great fluctuation of

porosity and permeability occurred only during Test 3. After the test, regardless of the

change in confining pressure, as indicated in Figure 3.25, they varied stably.

3.2.2 Experiment Series V: the Key Factors of Reconsolidation

Due to the effect of injected fluid and confining pressure on reconsolidation, Test 4 and

Test 5 were designed to investigate the key factors which played the role on

reconsolidation. More details are demonstrated and given in Table 3.5.

Table 3. 5: Experimental condition for Experiment Series V (Peng, 2009; Peng & Kovscek, 2011)

Test	Fracture		Injection	PH	Temperature	Confining Pressure
	Quantity	Direction				
4	2	direction of flow	neutral water	7	20 °C	400 psi
5	2	normal flow	buffer	10	200 °C	100 psi

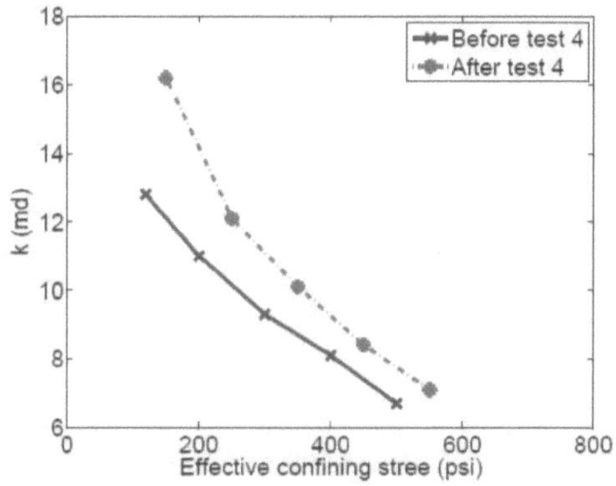

Figure 3. 26: Relationship between permeability and effective confining stress before and after Test 4 (Peng, 2009; Peng & Kovscek, 2011)

In Test 4, the neutral deionized water was injected into the matrix at room temperature

under confining pressure of 400 psi. Consequently, under these conditions the influence

of other factors was minimized. The permeability before and after Test 4 are shown in

Figure 3.26. Although the permeability changed slightly, the magnitude was the same.

The effect of confining pressure was minimized in Test 5 by having two fractures normal

to the injection direction. The variation in permeability in Figure 3.27 shows a slight difference from that in Figure 3.26, however, the difference was so small that it could be neglected.

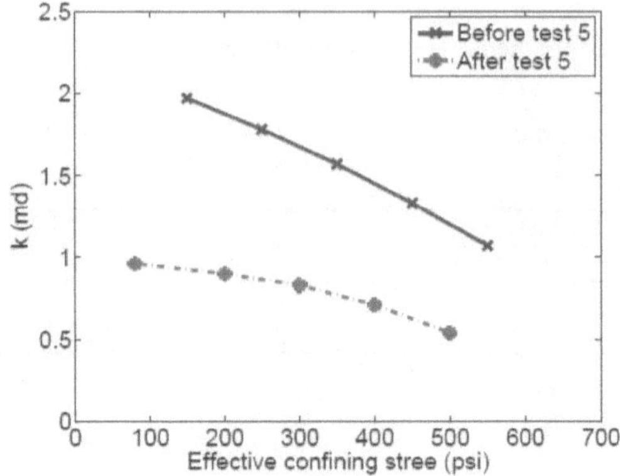

Figure 3. 27: Relationship between permeability and effective confining stress before and after Test 5 (Peng, 2009; Peng & Kovscek, 2011)

3.3   Discussion

3.3.1   Silica Dissolution

According to Reed, Rudenko and Sklyar, temperature, salinity, pH, and metal ions play a role in the process of dissolution (Reed, 1980; Rudenko & Sklyar, 1990). However, the influence of metal ions could be neglected (Peng, 2009; Peng & Kovscek, 2011). The process of dissolution in matrix during injection could be demonstrated with a set of reversible chemical formulas:

$$SiO_2(s) + 2H_2O \rightleftharpoons H_4SiO_4(aq) \qquad \text{Equation 3 1}$$

$$H_4SiO_4 + OH^- \rightleftharpoons H_3SiO_4^- + H_2O \qquad \text{Equation 3 2}$$

$$H_3SiO_4^- + OH^- \rightleftharpoons H_2SiO_4^{2-} + H_2O \qquad \text{Equation 3 3}$$

When the fluid was injected into the rock, silica was dissolved in aqueous solution to form silicic acid, shown as reversible Equation 3.1. Regardless of the pH of fluid, the process of dissolution was always promoted. Figure 3.1 could prove that this process even happened under acidic condition. However, alkaline solution seemed to be a key factor for dissolving, promoting the extent of dissolution further. It can be explained by the Equation 3.2 and 3.3. The hydroxide made the reversible reaction forwards. The greater the concentration of hydroxide, the more silica in the matrix was dissolved.

As suggested by Diabira, the dissolution is a slow process in matrix (Diabira et al., 2001). Parts of results from Experiment Series I, II, and III supported that there was no regular of silica dissolution when the amount of injected fluid was not enough.

The temperature had an effect on the process as well, increasing the kinetics of reaction. The higher the temperature, the more strenuous the dissolving was. Consequently, the two most important factors were pH and temperature, while the salinity as well as metal ions could be neglected.

3.3.2 The Process of Reconsolidation

As stated in the section of Experiment Series IV, based on results of Test 1, strong silica dissolution even happens under very acidic condition at elevated temperature. The enhanced rate constant at high temperature is the main reason. When Test 1 and Test 2 were compared, a similar extent of dissolution was found because of the same experimental temperature, which decided the kinetics of dissolving.

Furthermore, the injection was buffer solution in Test 2. The pH of the buffer solution was 10 before injection. Even though it was injected into the rock, the pH had only slightly fluctuated. It meant that this buffer solution with a stable value of pH was not affected by the silicic acid due to dissolution, which caused the greater extent of reaction. The increasing permeability proved the tendency.

In Test 3, reconsolidation was observed through the healing fracture. To confirm the factors, Tests 4 and 5 in Experiment Series V were conducted, which were intended to verify the influence of confining pressure and silica dissolution. Based on the results, either only confining pressure or only silica dissolution could not result in reconsolidation. Both conditions were necessary for reconsolidation.

The mechanism of reconsolidation could be interpreted by the theories of precipitation and gelation. Precipitation was believed to be a possible reason. However, the experimental conditions in Test 2 and 3, i.e. elevated temperature and alkaline pH, are favorable for silica dissolution. During the dissolution of silica, the precipitation also happened in the matrix as process is reversible. As one form of gelation, the generation of colloidal silica entails longer time. So, in this section the second form of gelation, silicate gelation would be discussed. Existing as an intermediate sate (Hench & West, 1990; RobertáHillman, 1996), silicate gelation happened in the studied pH range and the gelation time was relatively short. It was one important factor for the mechanism of fracture reconsolidation, but not the only one. Although others, such as pH, also had influence on precipitation, the confining pressure increased the rate of changing from silica gel to solid silica (Artaki, Sinha, Irwin, & Jonas, 1985).

To summarize, the process of reconsolidation takes place in three steps. Firstly, the silica in matrix should be dissolved. Then, under the definite conditions, especially the temperature and confining pressure, silica gelation happened. In the last step, confining pressure squeezed the fracture and made it smaller, which in turn enhanced the transformation of silica gel to solid silica.

4 Desorption

With the large-scale development and utilization of shale gas in the world, the knowledge about shale gas has changed recently (Tan et al., 2014). The focus on shale has transferred from absorption to desorption, especially in China. At the beginning of the boom of shale gas, lots of work was implemented, for instance, the research about modeling the mechanism of adsorption, the completely different flow regime in matrix studied by Foroozesh (Foroozesh, Abdalla, & Zhang, 2019), and the transport behavior was plotted by Bai (J. Bai et al., 2019), as shale previously identified as the cap rock for oil and gas reservoirs (Grunau, 1987; Schlömer & Krooss, 1997).

Until the Fuling shale gas field in the Sichuan has succeeded, which is the first commercial shale gas field in China (Dong et al., 2016; Yang et al., 2017), the question of how to increase the production was most frequently discussed. In correlation with increasing production, how to desorb more absorbed and stored gas in shale matrix should be comprehensively understood. Thus, it is meaningful to clarify the mechanism of desorption, and to find out the process of desorption and its related factors. Furthermore, understanding the mechanism and process of desorption could guide fracturing and help to exploit the resource more effectively. Based on the results in Chapter 2, tests relevant to desorption were conducted in this chapter. After that, the results were presented and demonstrated. In the last section, the factors which had an influence on desorption were analyzed and discussed. Additionally, the possible method to increase desorption was given too.

4.1 The Description of Desorption Test

To ensure that the desorption test was conducted under the in-situ condition, pressure coring was implemented to collect samples. The pressure coring technology can effectively prevent the shale gas from escaping during the process. All samples were selected from the Longmaxi Formation in the southeastern Sichuan Basin. The well logs in Figure 4.1 interpreted some characteristics of this formation.

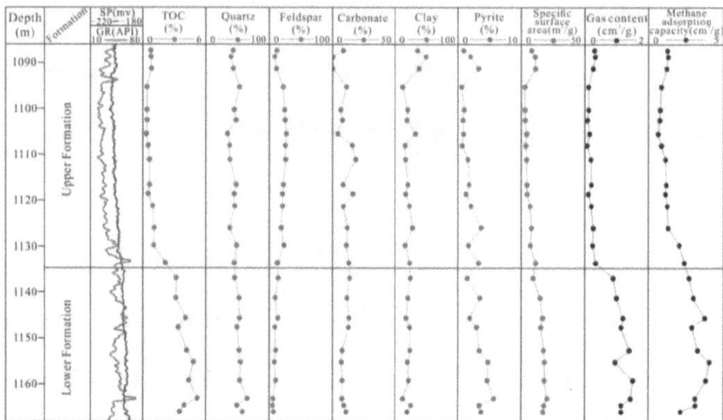

Figure 4. 1: Presentation of Longmaxi formation characteristics through well logs (Tang et al., 2019)

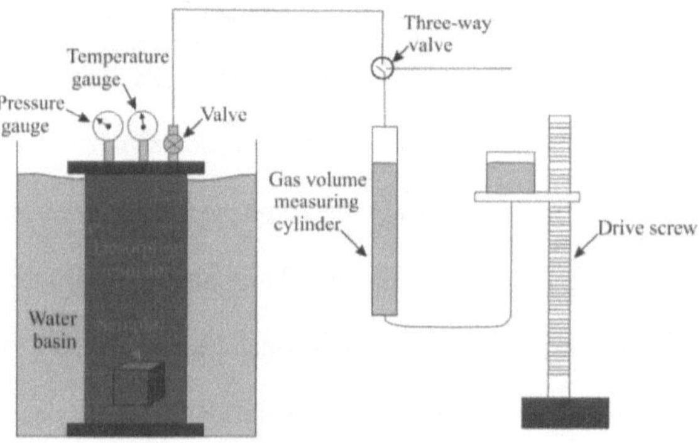

Figure 4. 2: The set-up of shale gas desorption tests (Tang et al., 2019)

In accordance to Tang (Tang et al., 2019), the experimental equipment consisted of desorption canister, water basin, gas valves, thermometer, pressure gauge, and measuring cylinder, as illustrated in Figure 4.2. The water bath was intended to control and keep the experimental temperature. The volume of desorbed gas was calculated by the displacement of acidified water. In this section the experimental temperatures were set at 50 °C and 98 °C. The former temperature represented the reservoir temperature, while the following was the highest value that could be reached in laboratory.

4.1.1 Results and Analysis

As presented in Figure 4.1, the formation could be divided into upper and lower formation. The gas content of the upper formation is less than that of the lower. With an average value of 0.20 m³/t, the shale gas content of upper reservoir varies from 0.07 m³/t - 0.35 m³/t, while the lower formation ranges from 0.90 m³/t - 1.52 m³/t, with an average value of 1.19 m³/t. The difference in the gas content between both parts of formation could be attributed to geological tectonic movement. The lower was formed in the deep water shelf facies with shale being rich in organic matter, while the upper reservoir was formed in the shallow water shelf facies with organic-poor shale (L. Chen et al., 2015; Guo, 2016). The associated information about total organic carbon (TOC), specific surface area, and methane adsorption capacity is listed in Table 4.1. As shown in Figure 4.1 and Figure 4.3, Longmaxi formation mainly consisted of quartz and feldspar, with an average fraction of 47%. Following are the clay minerals and carbonate minerals, with fractions of 22.9% and less than 20%, respectively.

4.2 Desorption Processes

25 cores had been used for desorption experiments by Tang (Tang et al., 2019). Samples 1 to 15, whose results are demonstrated in Figure 4.4, were derived from the upper formation with less gas content. Samples 16 to 25 have been taken from lower reservoir. Their results are shown in Figure 4.5. Both Figure 4.4 and Figure 4.5 described that the experimental samples underwent different desorption processes.

Table 4. 1: Comparison of TOC, specific surface area, and methane adsorption capacity in upper and lower formation defined in Figure 4.1 (Tang et al., 2019)

	Upper Formation	Lower Formation
TOC	0.75%	3.96%
Specific surface area	6.22 m^2/g	16.51 m^2/g
Methane adsorption capacity	0.98 m^3/t	3.36 m^3/t

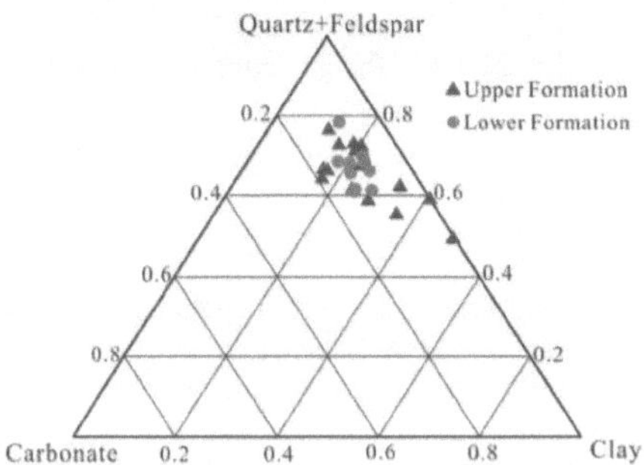

Figure 4. 3: Mineral compositions of shale from upper and lower formation (Tang et al., 2019)

Generally, at an experimental temperature of 50 °C, shown by the blue points in the figures, the trend of desorption was not pronounced. In contrast, a higher temperature exerts a greater influence on desorption. As a result, the red points denoting gas contents at 98 °C in Figure 4.4 and 4.5 were much more active than blue ones in most of the tests. Most of the curves have shown that samples had a much less loss at a lower temperature, in this section 50 °C compared to 98 °C. The average content of desorbed gas in the upper formation is about 31% at 50 °C. In comparison, 9% represented the average desorbed gas content of the lower formation.

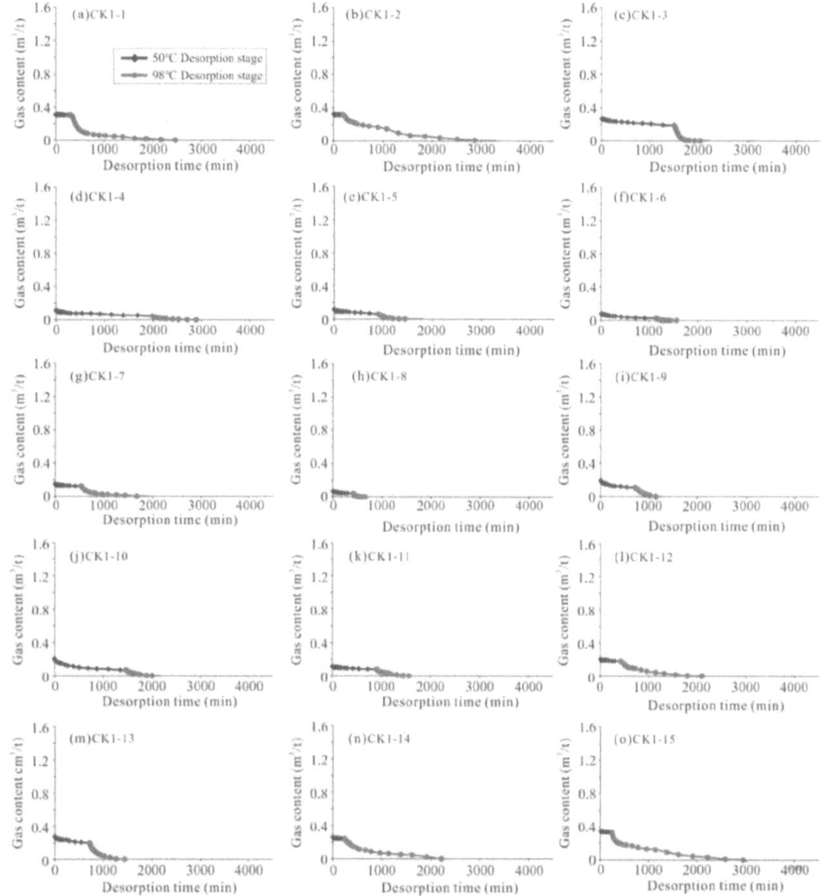

Figure 4. 4: Gas desorption of samples 1 to 15 from the upper formation (Tang et al., 2019)

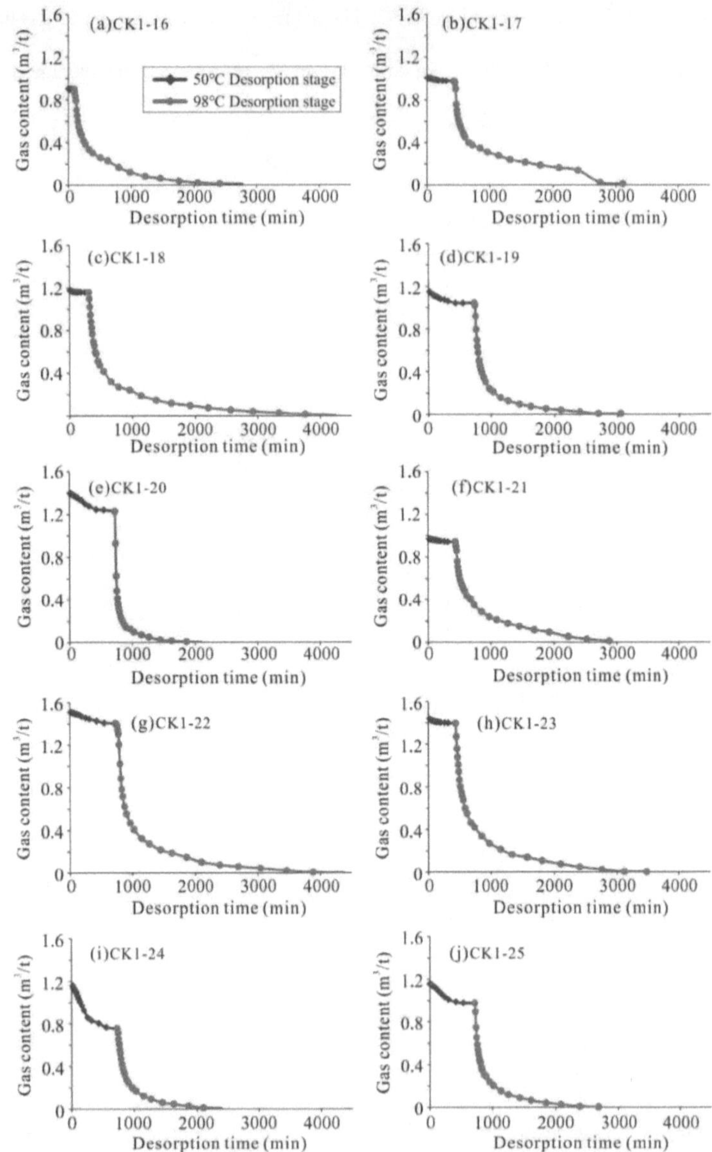

Figure 4. 5: Gas desorption of samples 16 to 25 from the lower formation (Tang et al., 2019)

Based on the statistics from laboratory, the most interesting facts had been found.
Although the gas content varied among the samples, some of the cores had the similar
desorption time. For instance, samples 1 and 23 had completely different gas content, but
about 80% of the gas desorbed within 1000 minutes in both cases.

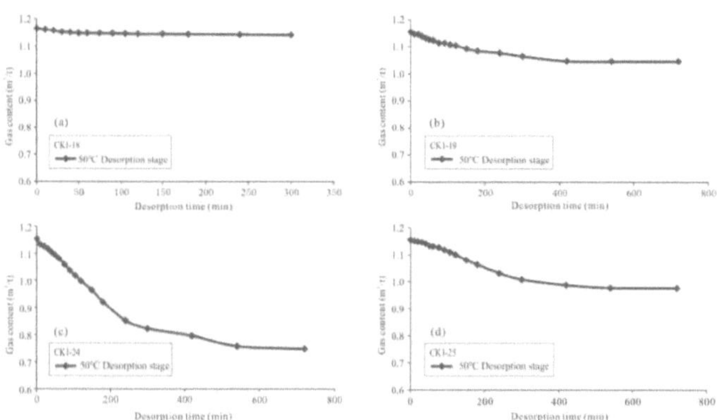

*Figure 4. 6: The desorption processes of samples 18, 19, 24, and 25 with similar gas
contents from the lower formation at 50 °C (Tang et al., 2019)*

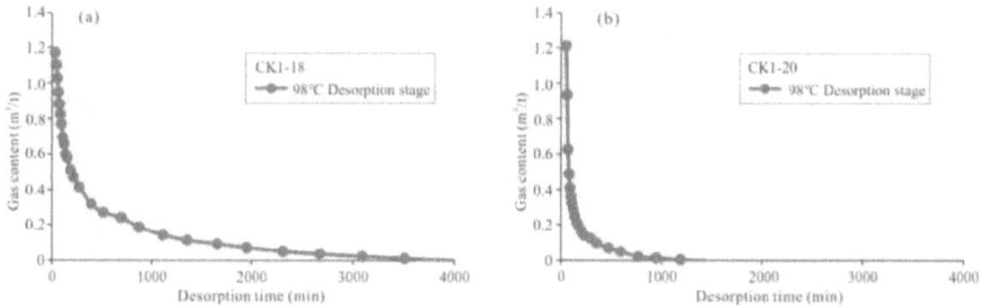

*Figure 4. 7: The desorption processes of samples 18 and 20 from the lower formation at
98 °C (Tang et al., 2019)*

Another interesting fact is that even though some samples possessed similar gas content,
their desorption curves exhibited clearly different trends. For example, shown in Figure
4.6, samples 18, 19, 24, and 25 had similar gas content, nearly 1.2 m^3/t. At a lower
temperature of 50 °C, sample 18 experienced only a slight gas loss, while sample 24 had
a heavy desorption, as demonstrated in Figure 4.6. Among the four samples, core 18 lost

the least gas, 0.02 m^3/t. Following were samples 19, 24, and 25, with the lost volume of 0.11, 0.41 and 0.18 m^3/t, respectively. At the beginning of process, sample 18 released gas slowly, while the desorption in core 24 occurred more rapidly and violently.

More specifically, there was difference in the velocity of desorption. For a better observation, the statistics of samples 18 and 20 at 98 °C are re-drawn in Figure 4.7. The desorption of gas took sample 18 around 4000 minutes, while for sample 20 it lasted approximately 1500 minutes. Furthermore, the slopes of the curves were significantly different. Number 18 had a slow desorption process, while core 20 lost 90% of the content in the first 200 minutes.

4.3　Discussion

4.3.1　Desorption Capability

The ability of desorption could be introduced firstly to analyze and explain the different performance of samples in the experiment of desorption process. In accordance with most of researches, the most important factors for desorption are temperature, pressure, and shale properties (Tang et al., 2019). As these parameters can be influenced by each other, their influences on shale properties could be demonstrated by those of the temperature as discussed below.

According to the equation of state, at a constant temperature of 98 °C, the relationship between pressure and gas content is a positive linear correlation (Tang et al., 2019). In addition, the shale desorption velocity is controlled by the shale gas content under the limitations of the experimental conditions (Tang, Jiang, Jiang, Cheng, & Zhang, 2017). As an example, the relationship between velocity and gas content of sample 23 is drawn in Figure 4.8. When the gas content decreased, the desorption velocity reduced.

Therefore, the desorption capability can be defined by a function of the velocity and gas content.

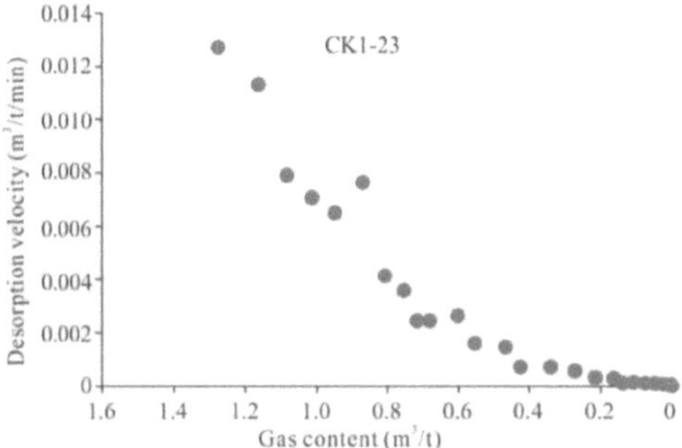

Figure 4. 8: The relationship between gas content and velocity of sample 23 at 98 °C (Tang et al., 2019)

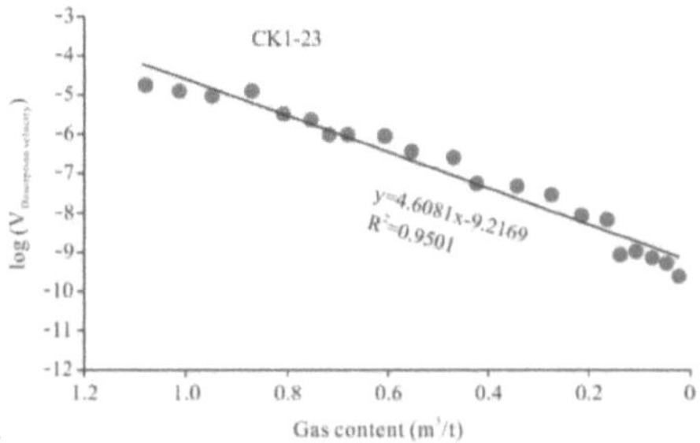

Figure 4. 9: The relationship between $\log(V_{desorption})$ and gas content (Tang et al., 2019)

As demonstrated by Tang (Tang et al., 2017; Tang et al., 2019), there was a good linear relationship between the logarithm of desorption velocity and gas content. For example, Figure 4.9 shows the logarithm of desorbed gas volume $\log(V_{desorption})$ as a function of initial gas content. The slope of the fitting curve represented the desorption capability, which is a quantitative value and is often regarded as a comprehensive parameter representing the shale properties.

Table 4.2 presented the desorption capacity of all samples at two different experimental temperatures. From the table, the desorption capacity of upper formation was generally greater than that of the lower. This is an important reason why the gas content of lower reservoir was higher than upper formation. Zhou also explained that Longmaxi formation is highly over mature shale, and no more gas is being generated, which led to desorption in the formation (Zhou, Xiao, Tian, & Pan, 2014).

4.3.2 The Percent of N_2

The shale gas compositions of the both upper and lower formation are listed in Table 4.3. It can be seen that most of the components have a greater content in the lower reservoir were greater than that in upper, such as CH_4. However, there were two significant differences which could not be neglected. One was feldspar, whose content in the upper formation was higher than in the lower. Another difference was the content of N_2. The statistics from laboratory presented that the average value of N_2 in the upper formation is higher than that in the lower formation, 9.96% and 3.37%, respectively.

Although the origin of N_2 in the shale is still controversial, according to the existed literature, there are some hypotheses, such as atmospheric exchange, mantle supple, and hydrocarbon source rock generation (Beaumont & Robert, 1999; Bräuer, Kämpf, Niedermann, Strauch, & Weise, 2004; Krooss et al., 1995). Due to the over mature property, it was impossible to regenerate more shale gas. And the content of nitrogen isotopes in the Lower Cambrian Niutitang Formation which is below the Longmaxi Formation is ranging from 2.6% to 0. It seemed that the exchange between shale and outside was responsible for the N_2 in the Longmaxi Formation.

Table 4. 2: Desorption capacity and proportions at 50 °C and 98 °C (Tang et al., 2019)

Format ion	Sam ple	50 °C		98 °C		Desorption capability at 98 °C	Correlation coefficient(R^2)
		Desorption content (m^3/t)	Desorption proportion (%)	Desorption content (m^3/t)	Desorption proportion (%)		
Upper	1	0.0061	1.97	0.3039	98.03	19.915	0.9046
	2	0.0055	1.72	0.3155	98.28	8.9957	0.7641
	3	0.078	29.22	0.189	70.78	9.9803	0.6666
	4	0.0716	62.28	0.0434	37.72	68.253	0.6863
	5	0.0606	50.09	0.0604	49.91	45.549	0.7544
	6	0.0548	66.84	0.0272	33.16	275.74	0.651
	7	0.0279	18.35	0.1241	81.65	41.838	0.816
	8	0.0286	43.29	0.0374	56.71	101.58	0.7688
	9	0.0848	43.51	0.1102	56.49	19.832	0.7125
	10	0.1386	66.31	0.0704	33.69	32.817	0.7646
	11	0.0335	30.19	0.0775	69.81	35.901	0.6261
	12	0.0258	12.63	0.1782	87.37	15.433	0.7207
	13	0.0782	28.54	0.1958	71.46	17.363	0.8969
	14	0.018	6.97	0.241	93.03	13.612	0.6465
	15	0.0133	3.81	0.3357	96.19	12.84	0.7862
Lower	16	0.0091	1.01	0.8939	98.99	6.7794	0.9598
	17	0.0324	3.23	0.9706	96.77	8.0119	0.9763
	18	0.0228	1.92	1.1632	98.08	6.462	0.9431
	19	0.1074	9.3	1.0466	90.7	7.7613	0.9444
	20	0.1713	12.19	1.2347	87.81	9.1802	0.9488
	21	0.0269	2.79	0.9381	97.21	6.3003	0.9756
	22	0.108	7.12	1.409	92.88	5.8831	0.9574
	23	0.0452	3.12	1.4038	96.88	4.6081	0.9501
	24	0.4067	35.24	0.7473	64.76	7.954	0.9842
	25	0.1797	15.54	0.9763	84.46	6.9513	0.9545

Table 4. 3: Shale gas composition (Tang et al., 2019)

Formation	Sample	Gas Composition (%)				
		CH_4	N_2	CO_2	C_2H_6	C_3H_8
Upper	1	93.75	4.5	0.93	0.81	0.01
	2	92.77	5.81	0.64	0.77	0.01
	3	92.09	6.96	0.35	0.59	0.01
	4	85.47	11.66	0.97	1.85	0.05
	5	93.42	5.3	0.62	0.65	0.01
	6	85.93	11.89	1.68	0.49	0.01
	7	89.19	8.91	1.13	0.75	0.01
	8	62.37	36.03	1.23	0.35	0.01
	9	89.53	7.84	1.46	1.14	0.03
	10	91.05	5.4	2.04	1.48	0.01
	11	84.99	13.67	0.73	0.61	0.01
	12	89.48	9.06	0.83	0.62	0.01
	13	94	4.36	0.96	0.67	0.01
	14	91.98	6.7	0.75	0.57	0.01
	15	87.44	11.31	0.6	0.64	0.01
Lower	16	95.32	3.6	0.47	0.61	0.01
	17	95.18	3.46	0.71	0.64	0.01
	18	95.02	3.95	0.43	0.59	0.01
	19	94.73	4.05	0.58	0.63	0.01
	20	96.08	2.85	0.41	0.65	0.01
	21	94.31	4.4	0.59	0.68	0.01
	22	96.28	2.66	0.41	0.64	0.01
	23	95.66	3.36	0.34	0.63	0.01
	24	94.87	2.8	1.12	1.2	0.02
	25	95.79	2.52	0.75	0.92	0.01

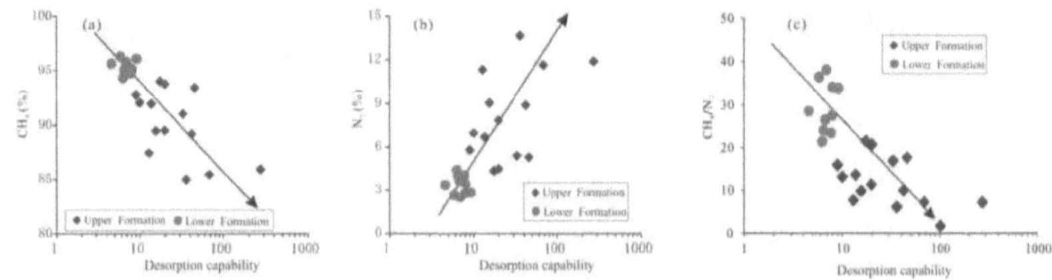

Figure 4. 10: The relationship between desorption capacity and N_2, CH_4, and CH_4/N_2 contents (Tang et al., 2019)

With the horizontal axis representing the desorption capability in Figure 4.10, it was easy

to identify that all the blue points located in the right side of red ones. That was, as

analyzed above in this section, the desorption of upper formation is greater than the lower. It could also be seen that the percentage of nitrogen in lower formation is less than the upper reservoir which has a relatively higher desorption capability. The probable reason might be that the N_2 could be translated into the shale, indicating that the shale has the capacity to receive external gas. If so, the higher the percentage of nitrogen is, the stronger the receiving (or exchange) capability of the shale is. This assumption might point a way for effective exploration of CH_4 in shale formation.

4.3.3   Feldspar

The percentage of feldspar was the second difference which could not be neglected in this section. The content could be retrieved from the well logs in Figure 4.1. Like the nitrogen, the content of feldspar in upper was much greater than in lower. Figure 4.11 shows that there was no obvious correlation between desorption capacity and carbonate or clay, compared with the relationship between the desorption capability and the feldspar. As plotted in Figure 4.11(a), except for the obvious difference in the feldspar content between upper and lower formation, the ability of desorption had a strong correlation with the content of feldspar. When the percentage of feldspar in shale risen, the desorption increased exponentially. The major reason was probably that feldspar developed pores well in the shale, as revealed by SEM image in Figure 4.12, which were much more easily for adsorbed and stored gas to migrate, lose, and exchange with the atmosphere. The more feldspar, the more pores well developed, and the more gas migrated and lost. However, feldspar had a small effect on methane adsorption (Tian, Li, Zhang, & Xiao, 2016). The slight influence on adsorption was attributed to the specific surface area, which will be introduced following part.

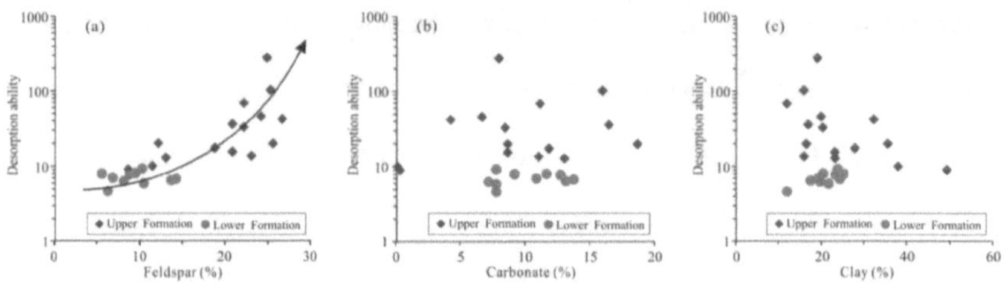

Figure 4. 11: The relationship between desorption capacity and (a) feldspar, (b) carbonate, and (c) clay (Tang et al., 2019)

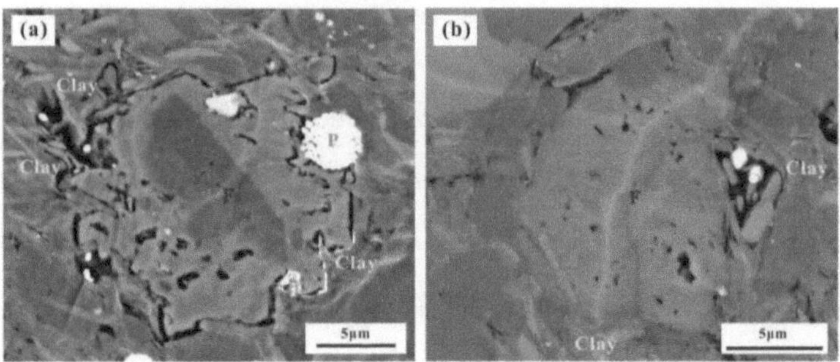

Figure 4. 12: Pores in feldspar (Tang et al., 2019)

4.3.4   Organic Matter

As proven by the statistics from Chapter 2 and Figure 4.1 of this chapter, shale consisted mainly of quartz, clay minerals, and organic matter. Although the percentage of organic matter is relatively low, varying from 0.5% to 5.1%, its influence on desorption is intense (Tang, Jiang, Huang, et al., 2016). Statistics revealed that the shale with greater shale gas storage had a relatively higher total organic carbon (TOC), regardless of another factor affecting on the gas content was the mature extent. In some extent, the percent of TOC could be a good indication. Figure 4.13 interpreted the relationship between TOC and desorption capability. Not surprisingly, TOC of upper and lower formations were very different from each other. With increasing total organic carbon content, the desorption capability decreased.

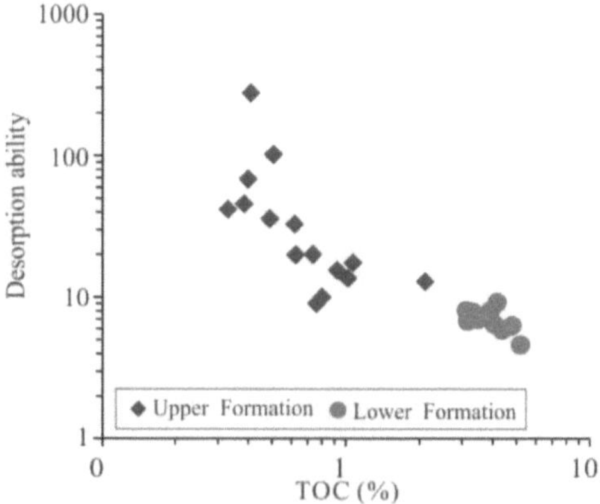

Figure 4. 13: Desorption capacity showing a negative correlation with the content of TOC (Tang et al., 2019)

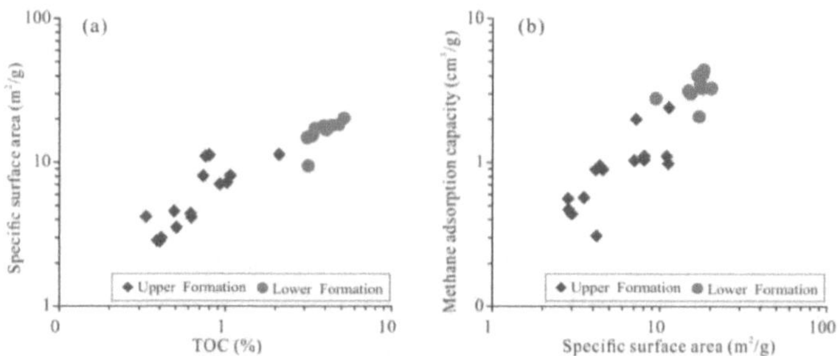

Figure 4. 14: (a) Specific surface area of shale as a function of its TOC content; (b) methane adsorption capacity as a function of specific surface area (Tang et al., 2019)

In order to comprehensively understand how the organic matter influenced the gas content, the surface specific area was introduced and analyzed. It is a property of solids defined as the total surface area of a material per unit of mass. Generally, the organic matter in shale has a high specific surface area, and a higher specific surface area is associated with a stronger adsorption capacity (Zhang, Ellis, Ruppel, Milliken, & Yang, 2012). Zhang has also reported that the content of TOC had positive correlation with the specific surface area, which would increase the property of absorption, as exemplified in Figure 4.14. It is therefore reasonable to assume that it would be difficult for the migration of CH_4, when the shale had a great specific surface area. Figure 4.15 illustrates that desorption capability exhibited a negative correlation with both the specific surface area and methane adsorption capacity. According to the Figure 4.16, the organic matter developed uncountable pores which had a complex structure inside shale matrix. In addition, those pores grew and merged when they met each other forming a complex pore network (Tang et al., 2015). As a consequence, the nanoscale pores had higher specific surface area which prevented the migration of shale gas.

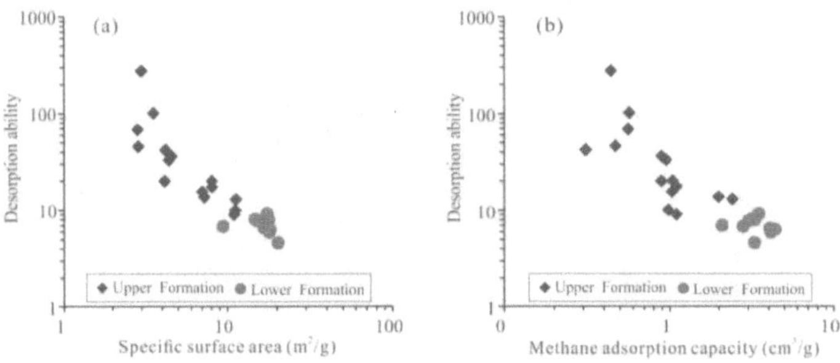

Figure 4. 15: Desorption ability showing negative correlations with both (a) specific surface area and (b) methane adsorption ability (Tang et al., 2019)

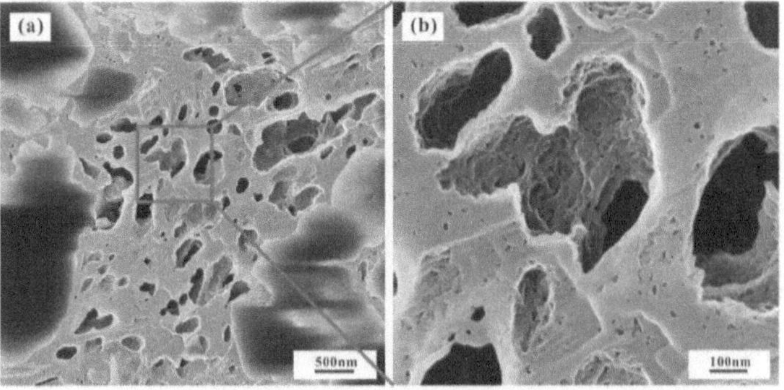

Figure 4. 16: Well-developed network of pores in organic matter, implying a high specific surface area (Tang et al., 2019)

4.3.5   Quartz and Clay Minerals

Compared to organic matter, quartz and clay minerals had a relatively smaller influence on desorption capability, although they were also the main component. Nevertheless, Figure 4.17 presented also a negative correlation between desorption capability and quartz or pyrite content. With the increase in quartz or pyrite content, the ability of desorption fell. The origin of reduced desorption ability was likely related to the organic matter in quartz, as this trend resembled the variation in desorption capability with increasing organic matter and quartz was biogenic and coupled to the organic matter (X. Liu, Xiong, & Liang, 2015).

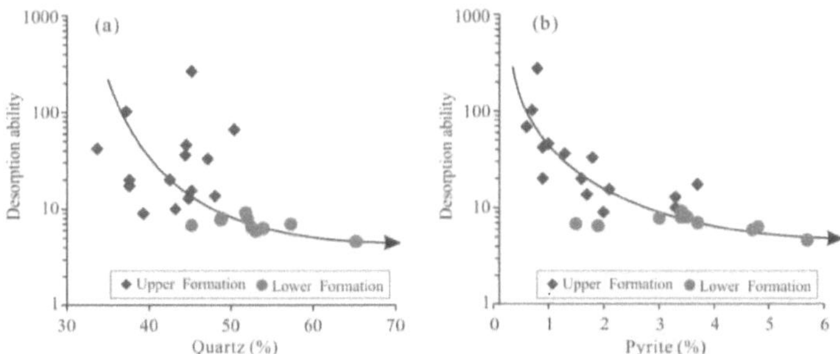

Figure 4. 17: Negative correlations between desorption ability and (a) quartz or (b) pyrite content (Tang et al., 2019)

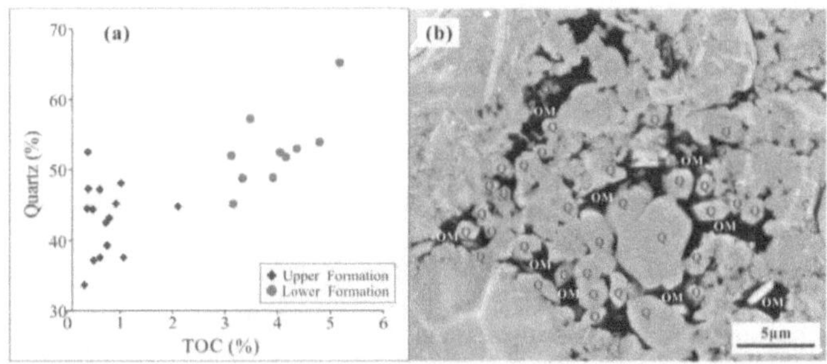

Figure 4. 18: (a) The content of TOC increased with the percentage of quartz; (b) the observation of SEM image (Tang et al., 2019)

Figure 4.18 showed that the quartz content is in a positive relationship with organic matter (OM) content. The content of quartz increased with a higher organic matter content. The biogenic quartz carried a lot of organic matter (Wenzhi, Jianzhong, Tao, Shufang, & Huang, 2016; Zhao, Jin, Jin, Wen, & Geng, 2017). The SEM image confirmed that there were only a few pores in quartz, resulting in difficulty in storing shale gas. The limitation of space in quartz resulted in a contribution to desorption, which could be neglected (Tang, Jiang, Jiang, & Li, 2016).

4.4   Conclusion

In this chapter, the difference between upper and lower formation were discussed and analyzed. The desorption capability was calculated as the slope of the $\log(V_{desorption})$ vs. initial gas content curve. Through the comparison, the upper reservoir did not only have much more gas content, but also have stronger ability of desorbing the shale gas.

This could be a result of many facts. One of them was that shale in the lower formation had more TOC because of the tectonic movement. The lower formation was formed in deep water with much more organic matter, while the upper was formed in shallow area which was organic poor.

As discussed above, the organic matter was the key factor. Apart from its influence on the reserved gas content, it played an important role in the ability of adsorption and desorption as well. The high organic matter content led to a complicated network of pores with high specific surface area and a strong methane adsorption capacity of the shale. And the network of pores also made the adsorbed gas difficult to migrate.

Compared with organic matter, the influence of clay minerals on desorption was only slight and could be neglected. The best evidence was the feldspar. The feldspar content could increase the shale gas desorption capability, as feldspar pores are large and well connected with each other, which is beneficial for the desorption and loss of shale gas. Although the pores in feldspar offered space for storing shale gas, it was also easy for gas to escape. In a word, the surface specific area was the best explanation for this obvious difference.

Other minerals, such as pyrite, carbonate, and clay, exerted no obvious effect on the shale gas desorption capability. However, quartz had a positive correlation with gas content and desorption capacity. This was because that organic matter developed inside quartz.

The most interesting finding was the exchange between N_2 and CH_4, which would benefit the exploration of production.

5 Optimization

As a global challenge, the extremely low permeability of shale plays has made it difficult to produce a commercially viable amount of gas (Wanniarachchi, Ranjith, & Perera, 2017). Therefore, appropriate advanced technologies are required. For example, shales with high quartz and low clay contents are brittle in nature (high Young's modulus and low Poisson's ratio), which had been described in Chapter 2. Therefore, some special techniques are required to enhance production capacity in shale formation, such as hydro-fracturing.

One of the main issues of hydro-fracturing is the sealing off of the generated fractures after releasing the applied pressure due to the high in situ stresses applied to the shale mass (Eia, 2011). As a result, proppants (such as sand, ceramics and resin-coated proppants) are normally injected with the fracturing fluid into the shale mass to prop open the created fractures after releasing the applied pressure (Wanniarachchi et al., 2017). In some shale plays, the proppants may disappear due to reconsolidation of proppants, which was introduced in Chapter 3.

According to Li, some gases have also been tested for the use as fracturing fluids particularly for shallow formations, for instance, CO_2 and N_2 (Xiang Li et al., 2015). Furthermore, gases like CO_2 and N_2 can be used in water-sensitive formations to avoid water interaction issues, and N_2 is especially useful to avoid and reduce swelling in shale clay minerals in the presence of water due to its potential for swelling recoverability (Gandossi, 2013). From Chapter 4, nitrogen has a great influence on shale gas production enhancement. In addition, according to the available experimental and modelling studies on foam fracturing, N_2-based foams are stronger than CO_2-based foams (Wanniarachchi et al., 2017).

Characterized by excellent proppant carrying capacity, quick flowing back, less formation damage and fluid loss (Gidley, 1989), foam fracturing technology that takes foam fluid as sand carrier is especially suitable for exploiting low permeable oil/gas reservoir and plays an important role in hydraulic fracturing (Sun, Liang, Wang, & Lu, 2014).

In this section, the foam fracturing fluid was chosen as an optimal option to increase the production of shale gas. It cannot only provide the basic function, such as fracturing and carrying proppants, but also solve the reconsolidation discussed in Chapter 3 through setting temperature and work with N_2 to exchange more CH_4 from matrix as indicated in Chapter 4. Due to the limitation of laboratory, only experiments regarding the choice of different foaming agents and stabilizers were conducted.

5.1 Proposal and Implementation

5.1.1 Proposal

The figures applied in previous chapters (Chapter 2, 3 and 4) were adopted from relevant literature after time consuming and intensive literature review studies. The main idea behind the literature review was to examine available fracture fluid and to come up with a suggestion how to optimize different factors and compositions of fracture fluid.
Therefore, the used figures have been interpreted independently and with support of well logging evidence, this led to investigate the idea of using N_2 or N_2-based fracture fluid. Own interpretations and many indications led to the conclusion that N_2 might be suitable in shale gas formation.

To summarize, the logic of optimization could be presented specifically. Based on the purpose of increasing the shale gas production, the characterization of shale formation has been described firstly. Quartz was relative sensitive under the condition of both temperature and pressure.

Due to thermodynamic behavior, the production capacity of CH_4 is proportional to the temperature. In contrary, the higher the temperature, the higher the acceleration of reconsolidation. Therefore, optimal pressure and temperature should be selected to optimize the production.

Feldspar and organic matter have a great influence on desorption, even though, they were still invariable. To change the content of feldspar or organic matter formed because of geological movement is impossible.

The phenomenon of exchanging N_2 for CH_4 offers the evidence that N_2-foam might be a good choice for the basic purpose, while foam could also adjust the temperature and pressure.

Therefore, it was decided to consider N_2-Foam as optimized frac fluid.

5.1.2   Implementation

To achieve the goals, the following procedures have been suggested:

- The implementation starts with the foaming tests. The several foam agents and additives should be selected for the optimal formulas. The agents and additives which can make uniform bubbles and longer half-life period are suitable for the further research.

- Afterwards, air-foam should be tested under different conditions to identify the reconsolidation. Most important operating parameters are temperature and pressure. In this section, various combinations consisted of different conditions should be evaluated in regard of foam stability. The combinations of different variations of temperature, pressure and foam which will not have an effect on the reconsolidation will be considered as an optimized condition. Foam formulation can be generated under the optimized conditions. To confirm exchange CH_4 by N_2, lab experiments can be conducted on shale samples.

- Other ability like transporting proppants should be considered after the lab experiments.

- Afterwards, pure nitrogen should be used to compare the results have obtained by air generated foam and the above procedures should be repeated. Pure N_2 is more effective than air; considering the additional costs of pure N_2, the air foam experiments have been conducted to compare the results for concluding the optimized suggestions.

Due to laboratory limitation, experiments regarding the choice of different foaming agents and stabilizers were conducted in this dissertation.

5.2 Foaming Test

Taking into account the utility and price, anionic foaming agent ABS and K_{12} were taken in these experiments. Their physical properties are listed in Table 5.1, and the parameters which had been measured for evaluating the foam quality are presented in Table 5.2. More parameters about foaming test had been measured and listed in Appendix A.

From Figure 5.1, when the mass fraction of the foaming agent was small ($\omega < 1‰$), surfactant molecules have not occupied the surface of a solvent, specifically water in this section, the surface tension decreased with agent fraction slightly, making it difficult to form a certain number of bubbles. With the increase in mass fraction ($1‰ < \omega < 3‰$), molecules of surfactant gradually filled the surface of water, which slowly reduced the surface tension of water and increasing the foam volume.

Table 5. 1: Physical properties of ABS and K_{12}

Type	Abbreviation	Name	Physical property
Anionic Surfactant	ABS	Sodium dodecyl benzene sulfonate	White or light yellow powdered solid, translucent shape after soluble in water, stable to the alkali, dilute acid and hard water, low surface tension, abundant in foam
	K_{12}	Sodium dodecyl sulfate	Light yellow powdered solid, translucent shape after soluble in water, stable to the alkali, dilute acid and hard water, strong foaming ability, and foam evenly

Table 5. 2: Main evaluation parameters

Parameter	Name	Description
V_F	Foaming volume	Under the speed of greater than 1000 r/min, the bubble volume in 100 ml base fluid produced by mixing after 1 min, ml
$t_{1/2}$	Half-life period	Required time to precipitate 50 ml liquid, min
$V_G : V_L$	Gas-liquid ratio	The ratio of foam volume to base fluid volume, dimensionless

As the gas, air in the experiment, was injected in the solution with surfactant, the bubbles were created and moved up. When the bubbles arrived to the surface of water, the surfactant molecules transferred themselves from the water surface to those bubbles forming a bilayer foam. In the latter stage of the foaming process, the quantity of surfactant molecules on the liquid surface gradually decreased, which in turn increased the surface tension, leading to the difficulty in further foaming. In other words, the ability to foam reduced gradually with time. At the end, the compact structure of foam lost and the size of bubbles were non-uniform.

When the mass fraction of surfactant kept increasing (3‰ < ω < 4‰), the number of molecules on the surface of solution was enough for foaming. The rest of surfactant molecules in water combined with each other, resulting in developing micelles in solution. Once the process of foaming started, as described above, the quantity of surfactant molecules decreased suddenly, while the micelles generated before decomposed themselves and supplemented the shortage on the surface of solution. As a result, the foam was generated continually, with much more uniform sizes and compact structure. However, due to the small amount, when the micelles were exhausted, the process of foaming slowed down. Meanwhile, the nonuniform foam appeared again unless the surfactant was added into the solution keeping the supplement to the loss on the water surface, for instance, ω > 4‰.

As an important parameter, half-life period serves a good reflection of the stability of foam. It shares a similar relationship with surfactant concentration as forming ability discussed above.

From Figure 5.1, if the concentration of the surfactant molecule is small (ω < 1‰), the surface tension of the solution is relatively strong, making it difficult to foam. When the molecules were enough for effectively reducing the surface tension of the solution (1‰ < ω < 3‰), a certain volume of foam was produced. However, the thickness varied non-uniformly as well as the structure did not show as compact because of the shortage of surfactant.

Laplace equation could be used to explain this phenomenon. According to the equation, pressure difference resulted in a decreasing thickness of bubbles. In addition, the small bubbles would be swallowed up by the larger ones, which led the larger bubbles to burst. At a higher concentration of surfactant molecule ($3‰ < \omega < 4‰$), a small number of micelles formed in the solution. Subsequently, uniform foam with compact structure formed, the impact of the pressure difference reduced, and the half-life period also extended. At an even higher mass fraction of surfactant ($\omega > 4‰$), the stable period extended further. This can be attributed the larger number of micelles consisting of surfactant molecules, which ensures the continuous formation of uniform bubbles because of the supplement of decomposing micelles. Finally, the influence of pressure and gravity reduced gradually, thereby stabilizing the foam.

Taking into account the volume of foam and half-life period K_{12} is a better agent than ABS as both parameters almost change simultaneously. As presented in Figure 5.1, curves of ABS began separating at the mass fraction of 4‰.

When the mass fraction was larger than 4‰, the foam volume continued rising with the increasing mass fraction. And stability also remained until the mass fraction exceeded 6‰. Although the half-life period remained high, its reduction could be observed from the statistics. The phenomenon probably suggested that the increase of ABS mass fraction simply reduced the surface tension of the solution, making it easier to foam. However, this easy foaming was associated with a better foam quality, specifically in terms of maintaining stability and extending the half-life period.

In contrast, the half-life period of foam generated by K_{12} was much longer than that by ABS after the further increase in mass fraction of K_{12} for improving the quality of the liquid film.

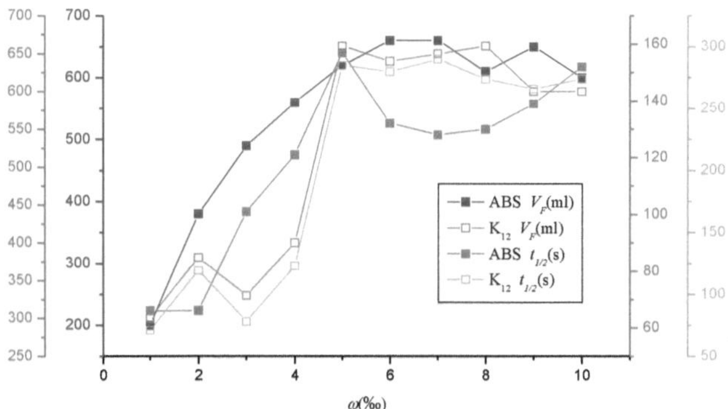

Figure 5. 1: Foam volume and half-life period comparisons between ABS and K_{12}

To summarize, both ABS and K_{12} give a similar maximum foam volume, although the associated half-life periods differed from each other. One of the possible reasons was that the non-uniform bubbles generated by ABS made it easy to form a differential pressure, which resulted in disappearance of the relatively small bubbles and reduction in the volume. Another convincing explanation was that the higher viscosity hindered the foaming process.

5.3   Experiment on Foam Stability

As interpreted in Figure 5.1, the maximum half-life period of foam generated by ABS or K_{12} was about 300 s with an unstable structure, as shown in Figure 5.2, which could not meet the requirements for the further experiments. Therefore, it was necessary to strengthen the film of foam and extend the stable time by adding stabilizers, as shown in Figure 5.3.

Figure 5. 2: ABS (left) and K_{12} foam

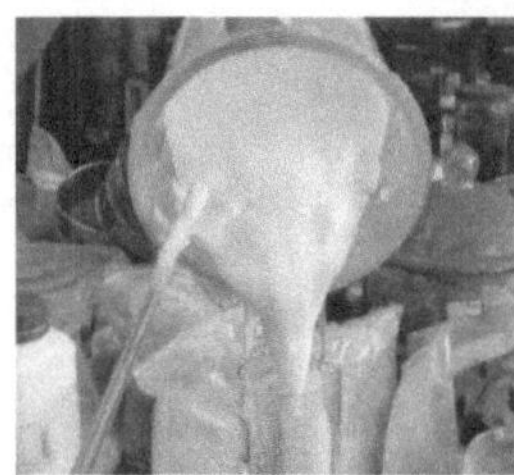

Figure 5. 3: Apparent increase in viscosity after adding the foam stabilizer

Experimental statistics in Figure 5.1 showed that half-life period of ABS foam became constant when the mass fraction exceeded about 4‰, while the foam volume had slight change after 6‰. The foam volume also changed slightly under the mass fraction of 4‰ and 6‰. Moreover, with similar half-life period, the optimal amount of foaming agent, 4‰ and 5‰ had been chosen for ABS and K_{12}, respectively.

Based on the consideration of practicality, seven stabilizers including HV-CMC, LV-PAC, HV-PAC, Sesbania, Konjac, XC and HEC, had been chosen for foam stability experiments. And the specific statistics of foam stability had been presented in Appendix B.

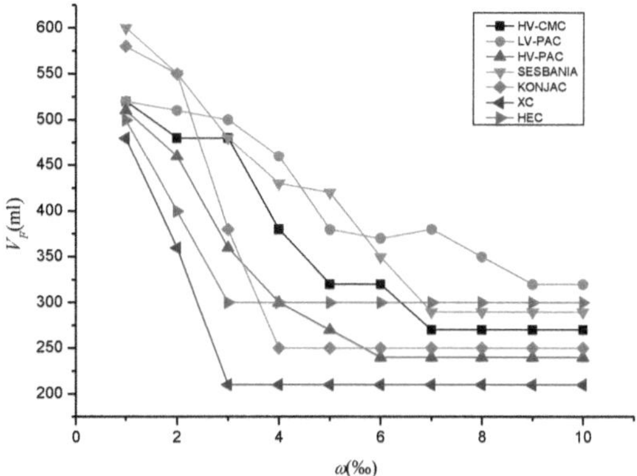

Figure 5. 4: Foam volume after adding the foam stabilizer into ABS

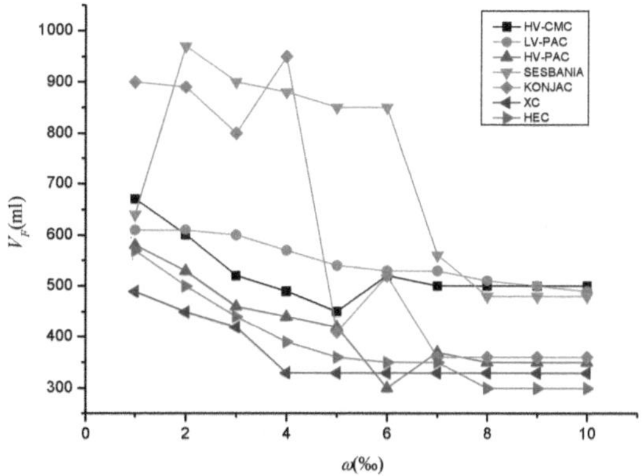

Figure 5. 5: Foam volume after adding foam stabilizer into K_{12}

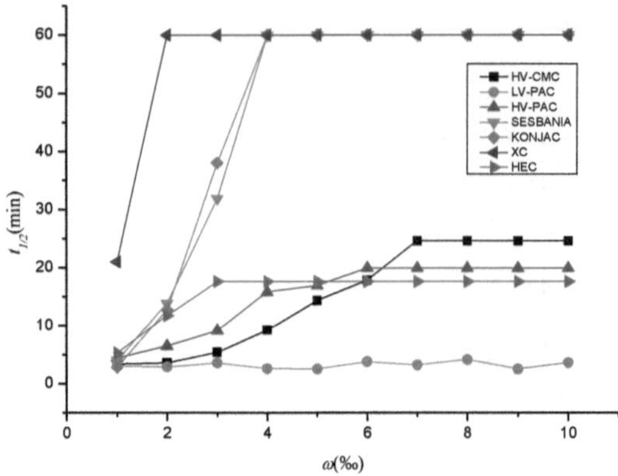

Figure 5. 6: The half-life period of ABS after adding foam stabilizer

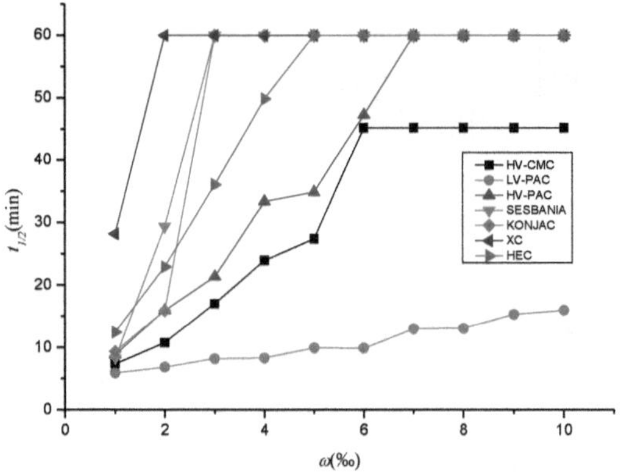

Figure 5. 7: The half-life period of K_{12} after adding foam stabilizer

Analysis from Figure 5.4 and 5.5 described that total volume of both ABS and K_{12} foam decreased with the increasing mass fraction. Moreover, ABS was greater than K_{12} in the term of amplitude of decrease when the content of stabilizer increased. The largest total foam volume appeared at the beginning of the plot. At a content of 1‰ added stabilizers, both Sesbania and Konjac, the generated foam volume was 600 ml. By contrast, the largest foam volume of K_{12} foam 950 ml, occurred at an addition of 2‰ Sesbania or 4‰ Konjac.

From Figure 5.4, the foam volume in seven experiments differed only slightly from each other at about 60 ml when the content varied from 1‰ to 4‰. In contrast, at the same range of K_{12} content the foam volumes after adding Sesbania and Konjac had a great difference with those after adding other stabilizers. In addition, the total volume decreased by 250 ml.

After foam stabilizer was added into the basic fluid containing the foaming agent, the viscosity rose up, which influences the foaming process and the final volume. The influence of stabilizer could be evaluated by the slope of V_F versus stabilizer w curved in Figures 5.4 to 5.7. The greater the slope was, the greater the effect on foaming was.

All added stabilizers had a greater influence on foaming of ABS. From Figure 5.4, the slope dropped abruptly. Compared with ABS, the situation from K_{12} was much more complex, especially with addition of Sesbania and Konjac.

After Sesbania was added into K_{12}, Figure 5.5, the total volume of foam increased firstly at 1 to 2‰ Sesbania. Then it kept constantly at 2‰ to 6‰ Sebania. With a further increase in the content of Sesbania, the foam volume reduced rapidly. In the term of Konjac, the value of foam volume fluctuated slightly within 1‰ to 4‰ Konjac. However, it decreased greatly at a Konjac mass fraction above 4‰. The value of volume was smaller than those with other five additives.

Figure 5.6 and Figure 5.7 demonstrated the relationship between the mass fraction of the foam stabilizer and half-life period. As shown, the increase in the mass fraction of the foam stabilizer resulted in the growth of the half-life period.

XC, Konjac, and Sesbania had greatest impact on the half-life period of ABS, despite adding seven foam stabilizers into the foaming agent. When the mass fraction of XC was 2‰, the half-life period had exceeded 60 min. For Konjac and Sesbania, the addition content should increase to 4‰ to reach the same half-life period of 60 min. The weakest influence appeared after LV-PAC was dissolved into basic fluid. It seemed that no effect was exerted on the length of duration. Regardless of the increasing stabilizer content, the half-life period remained around 3 min.

By contrast to ABS, the influence of different stabilizers which were dissolved into K_{12}, was more pronounced. The best performance still came from XC, Konjac and Sesbania. The longest half-life period of 60 min was achieved with only 2‰ of XC, 3‰ of Konjac, and 3‰ of Sesbania, respectively. HEC and HV-PVC additions required more percentage, 5‰ and 7‰ respectively to obtain the same half-life period. LV-PAC also had a relatively small effect on the duration of K_{12} foam, which fluctuated between the 5 min and the 15 min.

The slopes of $t_{1/2}$ vs. ω curves were in consistence with those of V_F vs. ω curves. The increasing viscosity was a convincing reason.

5.4   Further Experiments

According to the preceding discussion, it was easy to conclude that the influence of mono-additive was limited because of the viscosity. With the increase in the content of stabilizer, the foam volume decreased and the half-life period was longer. To weaken the effect which resulted from the viscosity of the mono-stabilizer, experiments with multi-additive were conducted and discussed in this section. Appendix C shown more details about the experimental process.

It can be concluded that the impact on ABS foam was weaker than that on K_{12} foam in terms of both the total foaming volume and the stability. In terms of additive, the HEC and Konjac had the ability to modify the parameter more than HV-CMC and HV-PAC. The final results of multi-additives were listed in Table 5.3 with three marked recommended formulas which could keep the balance between the increasing viscosity and decreasing foam volume. Figures 5.8, 5.9 and 5.10 interpreted the total volume and the half-life period using the recommended formulas.

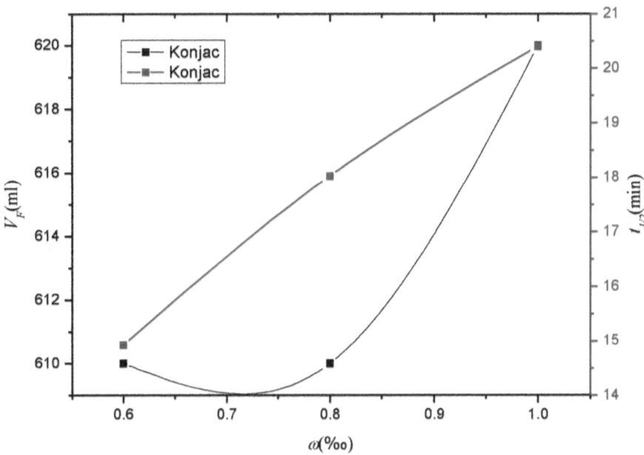

Figure 5. 8: 5‰ K_{12} + 0.6‰ Sesbania + Konjac

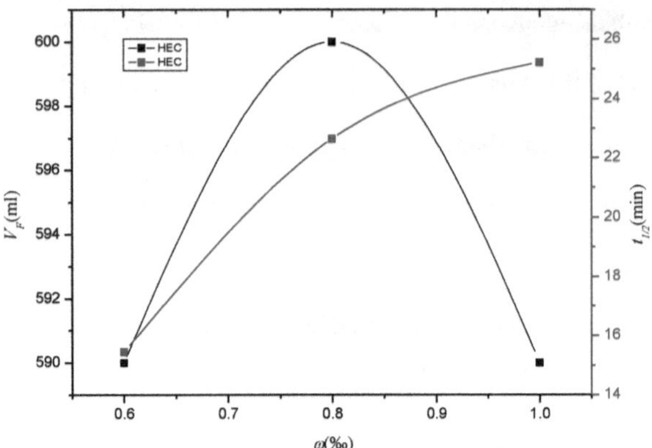

Figure 5. 9: 5‰ K_{12} + 0.8‰ Sesbania + HEC

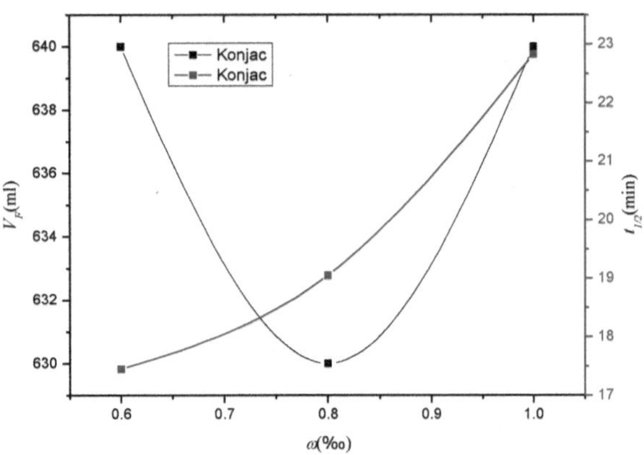

Figure 5. 10: 5‰ K_{12} + 0.8‰ Sesbania + Konjac

Table 5. 3: The experiments of multi-additives

Agents	Sesbania	2. Additive	Fluctuation		Maximum		Note
			V_F/ml	$t_{1/2}$/min	V_F/ml	$t_{1/2}$/min	
4‰ ABS	0.6‰	HV-CMC	30	4	520	9.5	
		HEC	40	6	490	15.5	
		HV-PAC	50	4	500	13	
		Konjac	20	6	480	15.5	
	0.8‰	HV-CMC	50	2.5	490	12.75	
		HEC	10	4.5	470	16	
		HV-PAC	10	2.5	460	12.25	
		Konjac	90	8	510	17	
	1.0‰	HV-CMC	40	4	460	16	
		HEC	0	4.5	410	20	
		HV-PAC	20	3	440	16	
		Konjac	30	22.5	440	7	
5‰ K$_{12}$	0.6‰	HV-CMC	0	2.25	625	12.25	
		HEC	50	9	600	24	
		HV-PAC	50	4.5	620	13.5	
		Konjac	10	6	620	20.5	recommended
	0.8‰	HV-CMC	50	4.5	650	18	
		HEC	10	10	600	25	recommended
		HV-PAC	10	4	610	19	
		Konjac	10	6	640	23	recommended
	1.0‰	HV-CMC	20	5.5	630	23.5	
		HEC	20	5.5	610	30.75	
		HV-PAC	30	6	610	25	
		Konjac	60	8.5	650	26	

6   Conclusions and Outlook

6.1   Conclusions

The purpose of this thesis is to increase the production of shale gas in China. Starting with two observations, reconsolidation and disappearance of proppants, the causes have been discussed in this dissertation on microscopic and macroscopic perspective. Experiments have been conducted in laboratory, mercury porosimetry and nitrogen gas adsorption sharing the distribution of pores in shale matrix. Combining with SEM and X-ray, more details and properties such as fracture and mineral content have been interpreted specifically. The further tests discovered the significant factors which caused the dissolution and reconsolidation inside the rock, for example, temperature and pressure. The statistics from the well logging supported the evidence of exchange between N_2 and CH_4, which makes an idea to strengthen exploration in field. As a possible suggestion, foam fracturing fluid has been introduced.

1. Micro- and mesopores dominated in shale matrix which consisted of two main parts, quartz and feldspar. Feldspar served a great space for reserving the gas, although it is easier to release. The dissolution of quartz made positive effect on fracturing, while the reverse reaction caused the reconsolidation.

2. Temperature influenced the process of both dissolution and reconsolidation. In contrast with steam, the liquid phase is suitable for the reaction of dissolving. In the meanwhile, the liquid condition is also positive for reconsolidation. Specially, both temperature and pressure are necessary for making reconsolidation happen inside matrix.

3. Organic matter supported specific surface area to trap shale gas in formation. The more specific surface area in the organic matter of matrix, the more absorbed gas in the formation. Although it is difficult to release absorbed gas directly, there is an exchange between N_2 and CH_4, which could be used for effective production.

4. By contrast with previous research, this thesis presents that clay minerals do not play a great role in absorption and desorption. Furthermore, the content of quartz could be an indicator to reveal where the reserve of the shale gas is rich. The organic matter always goes with quartz. The richer the content of quartz, the more organic matter, meaning more shale gas.

5. The foam fracturing fluid has been introduced as an optimal option to make the process of exploitation easier. ABS and K_{12} have been measured as foaming agents, which foams nitrogenous foam and unstable bubbles. Then seven additives have been mixed with those two agents for more uniform and stable foam. And in the next step three formulas including Sesbania and Konjac have been recommended.

6.2 Outlook

Theoretically, foam fluid could serve the significant functions which have positive effect on production. For instance, when the foam includes nitrogen, the advantage of exchange between N_2 and CH_4 could be achieved effectively. Meanwhile, the conditions which lead to reconsolidation could be avoided, such as causing less pressure. Based on the work performed in this thesis, further research could focus on the following points:

1. One of the most important factors effecting dissolution and reconsolidation, have not been implemented in this dissertation. The experiments should be designed and conducted under various temperature.

2. Afterwards, the pressure and proppants could also be considered as additional influence factors to simulate the formation environment.

3. N_2-foam is the core of this proposal. To verify the possibility of N_2-foam indispensable. Considering replacing air with pure nitrogen, the tests should be conducted repeatedly, comparing with "air group" for optimization.

4. Reasonably, the exchange mechanism between nitrogen and gas should be investigate.

References

Agency, I. E. (2009). *Energy Technology Perspectives 2010*.

Artaki, I., Sinha, S., Irwin, A., & Jonas, J. (1985). 29Si NMR study of the initial stage of the sol-gel process under high pressure. *Journal of non-crystalline solids, 72*(2-3), 391-402.

Bai, B., Elgmati, M., Zhang, H., & Wei, M. (2013). Rock characterization of Fayetteville shale gas plays. *Fuel, 105*, 645-652. doi:https://doi.org/10.1016/j.fuel.2012.09.043

Bai, J., Kang, Y., Chen, M., Liang, L., You, L., & Li, X. (2019). Investigation of multi-gas transport behavior in shales via a pressure pulse method. *Chemical Engineering Journal, 360*, 1667-1677. doi:https://doi.org/10.1016/j.cej.2018.10.197

Beaumont, V., & Robert, F. (1999). Nitrogen isotope ratios of kerogens in Precambrian cherts: a record of the evolution of atmosphere chemistry? *Precambrian Research, 96*(1-2), 63-82.

Bell, E., Kramer, S., Zajanc, D., & Aspittle, J. (2008). Salmonid Fry Stranding Mortality Associated with Daily Water Level Fluctuations in Trail Bridge Reservoir, Oregon. *North American Journal of Fisheries Management, 28*(5), 1515-1528. doi:10.1577/M07-026.1

Bhat, S., & Kovscek, A. (1999). Statistical network theory of silica deposition and dissolution in diatomite. *In Situ, 23*(1), 21-53.

Bräuer, K., Kämpf, H., Niedermann, S., Strauch, G., & Weise, S. M. (2004). Evidence for a nitrogen flux directly derived from the European subcontinental mantle in the Western Eger Rift, central Europe. *Geochimica et cosmochimica acta, 68*(23), 4935-4947.

Brunauer, S., Emmett, P. H., & Teller, E. (1938). Adsorption of gases in multimolecular layers. *Journal of the American chemical society, 60*(2), 309-319.

Bustin, R. M., Bustin, A. M., Cui, A., Ross, D., & Pathi, V. M. (2008). *Impact of shale properties on pore structure and storage characteristics.* Paper presented at the SPE shale gas production conference.

Chen, L., Lu, Y., Jiang, S., Li, J., Guo, T., Luo, C., & Xing, F. (2015). Sequence stratigraphy and its application in marine shale gas exploration: A case study of the Lower Silurian Longmaxi Formation in the Jiaoshiba shale gas field and its adjacent area in southeast Sichuan Basin, SW China. *Journal of Natural Gas Science and Engineering, 27*, 410-423.

Chen, S., Zhu, Y., Wang, H., Liu, H., Wei, W., & Fang, J. (2011). Shale gas reservoir characterisation: A typical case in the southern Sichuan Basin of China. *Energy, 36*(11), 6609-6616. doi:https://doi.org/10.1016/j.energy.2011.09.001

Diabira, I., Castanier, L., & Kovscek, A. (2001). Porosity and permeability evolution accompanying hot fluid injection into diatomite. *Petroleum science and technology, 19*(9-10), 1167-1185.

Dong, D., Wang, Y., Li, X., Zou, C., Guan, Q., Zhang, C., . . . Liu, H. (2016). Breakthrough and prospect of shale gas exploration and development in China. *Natural Gas Industry B, 3*(1), 12-26.

Dudley, B. (2018). BP statistical review of world energy. *BP Statistical Review, London, UK, accessed Aug, 6*, 2018.

Dullien, F., & Dhawan, G. (1974). Characterization of pore structure by a combination of quantitative photomicrography and mercury porosimetry. *Journal of Colloid and Interface Science, 47*(2), 337-349.

Eia, U. (2011). Review of emerging resources: US Shale gas and shale oil plays. *Energy Information Administration, US Department of Energy*.

El-Sayed, M., El Domiaty, A., I Mourad, A.-H., & Lotfy, M. (2018). *Leak Before Break Fracture Assessment for X70 Steel Pressurized Tubes With Axial Surface Crack.* Paper presented at the ASME 2018 Pressure Vessels and Piping Conference.

Fang, C. (2017). *Laboratory and Modeling Study of Fracability in Lacustrine Gas Shale of Ordos Basin, China.* Technische Universität Bergakademie Freiberg,

Fang, C., & Amro, M. (2014). *Pore structure characteristics of non-marine shale in Ordos Basin, China.* Paper presented at the IPTC 2014: International Petroleum Technology Conference.

Fang, C., Amro, M., Jiang, G., & Lu, H. (2016). Laboratory studies of non-marine shale porosity characterization. *Journal of Natural Gas Science and Engineering, 33*, 1181-1189. doi:https://doi.org/10.1016/j.jngse.2016.04.006

Fjar, E., Holt, R. M., Raaen, A., Risnes, R., & Horsrud, P. (2008). *Petroleum related rock mechanics* (Vol. 53): Elsevier.

Foroozesh, J., Abdalla, A. I. M., & Zhang, Z. (2019). Pore Network Modeling of Shale Gas Reservoirs: Gas Desorption and Slip Flow Effects. *Transport in Porous Media, 126*(3), 633-653. doi:10.1007/s11242-018-1147-6

Gandossi, L. (2013). An overview of hydraulic fracturing and other formation stimulation technologies for shale gas production. *Eur. Commisison Jt. Res. Cent. Tech. Reports, 26347.*

Geny, F. (2010). *Can Unconventional Gas be a Game Changer in European Gas Markets?* OIES paper: NG 46. Oxford Institute for Energy Studies.

Gidley, J. L. (1989). Recent advances in hydraulic fracturing.

Global, B. P. (2010). BP statistical review of world energy. *London, 64th Edition, BP Statistical.*

Grunau, H. R. (1987). A worldwide look at the cap-rock problem. *Journal of Petroleum Geology, 10*(3), 245-265.

Guo, T. (2016). Key geological issues and main controls on accumulation and enrichment of Chinese shale gas. *Petroleum Exploration and Development, 43*(3), 317-326.

Haifang, S. (2017). Xie Yi. Cleaner production in the Changning–Weiyuan National Shale Gas Demonstration Area. *Natural Gas Industry, 37*(1), 105-111.

Hench, L. L., & West, J. K. (1990). The sol-gel process. *Chemical reviews, 90*(1), 33-72.

Hoefner, M. L., & Fogler, H. S. (1988). Pore evolution and channel formation during flow and reaction in porous media. *AIChE Journal, 34*(1), 45-54. doi:10.1002/aic.690340107

Hou, Z., Xie, H., Zhou, H., Were, P., & Kolditz, O. (2015). Unconventional gas resources in China. *Environmental Earth Sciences, 73*(10), 5785-5789. doi:10.1007/s12665-015-4393-8

Krooss, B., Littke, R., Müller, B., Frielingsdorf, J., Schwochau, K., & Idiz, E. (1995). Generation of nitrogen and methane from sedimentary organic matter: implications on the dynamics of natural gas accumulations. *Chemical Geology, 126*(3-4), 291-318.

Kundert, D. P., & Mullen, M. J. (2009). *Proper Evaluation of Shale Gas Reservoirs Leads to a More Effective Hydraulic-Fracture Stimulation.* Paper presented at the SPE Rocky Mountain Petroleum Technology Conference, Denver, Colorado. https://www.onepetro.org:443/download/conference-paper/SPE-123586-MS?id=conference-paper%2FSPE-123586-MS

Li, X., Feng, Z., Han, G., Elsworth, D., Marone, C., & Saffer, D. (2015). *Hydraulic fracturing in shale with H 2 O, CO 2 and N 2.* Paper presented at the 49th US Rock Mechanics/Geomechanics Symposium.

Li, X., Mao, H., Ma, Y., Wang, B., Liu, W., & Xu, W. (2020). Life cycle greenhouse gas emissions of China shale gas. *Resources, Conservation and Recycling, 152*, 104518. doi:https://doi.org/10.1016/j.resconrec.2019.104518

Liu, X., Xiong, J., & Liang, L. (2015). Investigation of pore structure and fractal characteristics of organic-rich Yanchang formation shale in central China by nitrogen adsorption/desorption analysis. *Journal of Natural Gas Science and Engineering, 22*, 62-72.

Liu, Y., & Moh'd M, A. (2018). Modeling Shale Rock Behavior Using a Flat Joint Model. *Electronic Journal of Geotechnical Engineering, 23*, 611-623.

Loucks, R. G., Reed, R. M., Ruppel, S. C., & Jarvie, D. M. (2009). Morphology, genesis, and distribution of nanometer-scale pores in siliceous mudstones of the Mississippian Barnett Shale. *Journal of sedimentary research, 79*(12), 848-861.

Mason, G. (1982). The effect of pore space connectivity on the hysteresis of capillary condensation in adsorption—desorption isotherms. *Journal of Colloid and Interface Science, 88*(1), 36-46.

Montgomery, S. L., Jarvie, D. M., Bowker, K. A., & Pollastro, R. M. (2005). Mississippian Barnett Shale, Fort Worth basin, north-central Texas: Gas-shale play with multi-trillion cubic foot potential. *American Association of Petroleum Geologists Bulletin, 89*(2), 155-175. doi:10.1306/09170404042

Müller, M. (2017). *Laborative und mathematisch-numerische Untersuchung und Bewertung der Durchlässigkeit von Fließwegen bei der Stimulation von Sonden in Fluidlagerstätten unter besonderer Berücksichtigung des machanischen Kontaktes zwischen Proppants und Formation.* Technische Universität Bergakademie Freiberg,

Nobuyuki, H. (2009). Natural Gas in China: Market evolution and strategy. *International Energy Agency, Working papers Series.–2009 [Електронний ресурс].–Режим доступу: http://www. iea. org/Papers/2009/nat_gas_china. pdf.*

Peng, J. (2009). *The Impact of Oil Chemistry on Heavy-Oil Solution Gas Drive and Fracture Reconsolidation of Diatomite During Thermal Operations.* Stanford University,

Peng, J., & Kovscek, A. R. (2011). Temperature-Induced Fracture Reconsolidation of Diatomaceous Rock During Forced Water Imbibition. *SPE Reservoir Evaluation & Engineering, 14*(01), 45-59. doi:10.2118/132411-PA

Qu, Y., Tian, Y., Zou, B., Zhang, J., Zheng, Y., Wang, L., . . . Wang, Z. (2010). A novel mesoporous lignin/silica hybrid from rice husk produced by a sol–gel method. *Bioresource Technology, 101*(21), 8402-8405. doi:https://doi.org/10.1016/j.biortech.2010.05.067

Reed, M. (1980). Gravel pack and formation sandstone dissolution during steam injection. *Journal of Petroleum Technology, 32*(06), 941-949.

RobertáHillman, A. (1996). Crystal impedance: a new technique for monitoring the sol–gel process. *Journal of the Chemical Society, Faraday Transactions, 92*(6), 1079-1082.

Rudenko, L., & Sklyar, V. (1990). Bibliographic information regulation of the properties of diatomite. III. Effect of metal hydroxides and chlorides on permeability. *Izvestiya vuzov. Khimiya I Khimicheskaya Tekhnologiya, 33*(2), 27-30.

Schlömer, S., & Krooss, B. (1997). Experimental characterisation of the hydrocarbon sealing efficiency of cap rocks. *Marine and Petroleum Geology, 14*(5), 565-580.

Sing, K., Everett, D., Haul, R., Moscou, L., Pierotti, R., Rouquérol, J., & Siemieniewska, T. (1985). IUPAC Commission on colloid and surface chemistry including catalysis. Reporting physisorption data for gas/solid systems with special reference to the determination of surface area and porosity (Recommendations 1984). *Pure Appl Chem, 57*, 603-619.

Speight, J. G. (2013a). Chapter 1 - Origin of Shale Gas. In J. G. Speight (Ed.), *Shale Gas Production Processes* (pp. 1-23). Boston: Gulf Professional Publishing.

Speight, J. G. (2013b). Chapter 2 - Shale Gas Resources. In J. G. Speight (Ed.), *Shale Gas Production Processes* (pp. 25-68). Boston: Gulf Professional Publishing.

Speight, J. G. (2013c). Chapter 3 - Production Technology. In J. G. Speight (Ed.), *Shale Gas Production Processes* (pp. 69-100). Boston: Gulf Professional Publishing.

Speight, J. G. (2014). *The chemistry and technology of petroleum*: CRC press.

Speight, J. G. (2016). 1 - Introduction to fuel flexible energy. In J. Oakey (Ed.), *Fuel Flexible Energy Generation* (pp. 3-27). Boston: Woodhead Publishing.

Sun, X., Liang, X., Wang, S., & Lu, Y. (2014). Experimental study on the rheology of CO2 viscoelastic surfactant foam fracturing fluid. *Journal of Petroleum Science and Engineering, 119*, 104-111. doi:https://doi.org/10.1016/j.petrol.2014.04.017

Tan, J., Weniger, P., Krooss, B., Merkel, A., Horsfield, B., Zhang, J., . . . Tocher, B. A. (2014). Shale gas potential of the major marine shale formations in the Upper Yangtze Platform, South China, Part II: Methane sorption capacity. *Fuel, 129*, 204-218.

Tang, X., Jiang, Z., Huang, H., Jiang, S., Yang, L., Xiong, F., . . . Feng, J. (2016). Lithofacies characteristics and its effect on gas storage of the Silurian Longmaxi marine shale in the southeast Sichuan Basin, China. *Journal of Natural Gas Science and Engineering, 28*, 338-346.

Tang, X., Jiang, Z., Jiang, S., Cheng, L., & Zhang, Y. (2017). Characteristics and origin of in-situ gas desorption of the Cambrian Shuijingtuo Formation shale gas reservoir in the Sichuan Basin, China. *Fuel, 187*, 285-295.

Tang, X., Jiang, Z., Jiang, S., Cheng, L., Zhong, N., Tang, L., . . . Zhou, W. (2019). Characteristics, capability, and origin of shale gas desorption of the Longmaxi Formation in the southeastern Sichuan Basin, China. *Scientific Reports, 9*(1), 1035. doi:10.1038/s41598-018-37782-2

Tang, X., Jiang, Z., Jiang, S., & Li, Z. (2016). Heterogeneous nanoporosity of the Silurian Longmaxi Formation shale gas reservoir in the Sichuan Basin using the QEMSCAN, FIB-SEM, and nano-CT methods. *Marine and Petroleum Geology, 78*, 99-109.

Tang, X., Jiang, Z., Li, Z., Gao, Z., Bai, Y., Zhao, S., & Feng, J. (2015). The effect of the variation in material composition on the heterogeneous pore structure of high-maturity shale of the Silurian Longmaxi formation in the southeastern Sichuan Basin, China. *Journal of Natural Gas Science and Engineering, 23*, 464-473.

Tian, H., Li, T., Zhang, T., & Xiao, X. (2016). Characterization of methane adsorption on overmature Lower Silurian–Upper Ordovician shales in Sichuan Basin, southwest China: Experimental results and geological implications. *International Journal of Coal Geology, 156*, 36-49.

Wanniarachchi, W. A. M., Ranjith, P. G., & Perera, M. S. A. (2017). Shale gas fracturing using foam-based fracturing fluid: a review. *Environmental Earth Sciences, 76*(2), 91. doi:10.1007/s12665-017-6399-x

Washburn, E. W. (1921). The Dynamics of Capillary Flow. *Physical Review, 17*(3), 273-283. doi:10.1103/PhysRev.17.273

Wei, Y.-M., & Liao, H. (2016). *Energy economics: energy efficiency in China*: Springer.

Wenzhi, Z., Jianzhong, L., Tao, Y., Shufang, W., & Huang, J. (2016). Geological difference and its significance of marine shale gases in South China. *Petroleum Exploration and Development, 43*(4), 547-559.

Wu, P., & Aguilera, R. (2012). *Investigation of Gas Shales at Nanoscale Using Scan Electron Microscopy, Transmission Electron Microscopy and Atomic Force Microscopy*. Paper presented at the SPE Annual Technical Conference and Exhibition, San Antonio, Texas, USA. https://doi.org/10.2118/159887-MS

Yang, R., He, S., Hu, Q., Sun, M., Hu, D., & Yi, J. (2017). Applying SANS technique to characterize nano-scale pore structure of Longmaxi shale, Sichuan Basin (China). *Fuel, 197*, 91-99.

Zhang, T., Ellis, G. S., Ruppel, S. C., Milliken, K., & Yang, R. (2012). Effect of organic-matter type and thermal maturity on methane adsorption in shale-gas systems. *Organic geochemistry, 47*, 120-131.

Zhao, J., Jin, Z., Jin, Z., Wen, X., & Geng, Y. (2017). Origin of authigenic quartz in organic-rich shales of the Wufeng and Longmaxi Formations in the Sichuan Basin, South China: Implications for pore evolution. *Journal of Natural Gas Science and Engineering, 38*, 21-38.

Zhou, Q., Xiao, X., Tian, H., & Pan, L. (2014). Modeling free gas content of the Lower Paleozoic shales in the Weiyuan area of the Sichuan Basin, China. *Marine and Petroleum Geology, 56*, 87-96.

Appendix A

Appendix A. 1: Foaming test Nr. 1

Agents	Content	Measurements							Calculations		
		V_F	$t_{1/2}$	t_s	m_{50}	t_l	t_f	t_e	ρ_l	ρ_f	$V_G : V_L$
ABS	1‰	220	-	-	121.77	0s84	-	-	0.9714	0.4415	2.2
	2‰	290	-	-	122.35	0s85	-	-	0.9830	0.3340	2.9
	3‰	470	1m40s	-	122.71	-	-	-	0.9902	0.2107	4.7
	4‰	510	1m45s	-	122.13	-	-	-	0.9786	0.1919	5.1
	5‰	620	2m37s	-	122.61	-	-	-	0.9882	0.1594	6.2
	6‰	550	2m21s	-	122.03	-	-	-	0.9766	0.1776	5.5
	7‰	630	2m20s	-	122.37	-	-	-	0.9834	0.1561	6.3
	8‰	610	2m10s	-	122.45	-	-	-	0.9850	0.1615	6.1
	9‰	560	1m58s	-	121.84	-	-	-	0.9728	0.1737	5.6
	10‰	550	2m	-	122.64	-	-	-	0.9888	0.1798	5.5
K$_{12}$	1‰	310	1m17s	-	121.98	-	-	-	0.9756	0.3147	3.1
	2‰	380	1m59s	-	122.15	-	-	-	0.9790	0.2576	3.8
	3‰	350	1m23s	-	122.17	-	-	-	0.9794	0.2798	3.5
	4‰	400	2m3s	-	121.98	-	-	-	0.9756	0.2439	4.0
	5‰	660	4m45s	-	122.64	-	-	-	0.9888	0.1498	6.6
	6‰	640	4m40s	-	122.51	-	-	-	0.9862	0.1541	6.4
	7‰	650	4m50s	-	122.65	-	-	-	0.9890	0.1522	6.5
	8‰	660	4m34s	-	122.45	-	-	-	0.9850	0.1492	6.6
	9‰	600	4m11s	-	122.10	-	-	-	0.9780	0.1630	6.0
	10‰	600	4m23s	-	121.78	-	-	-	0.9716	0.1619	6.0

Appendix A. 2: Foaming test Nr. 2

Agents	Content	Measurements							Calculations		
		V_F	$t_{1/2}$	t_s	m_{50}	t_l	t_f	t_e	ρ_l	ρ_f	$V_G : V_L$
ABS	1‰	190	-	-	121.43	1s	-	-	0.9646	0.5077	1.9
	2‰	380	1m6s	-	122.51	0s81	-	-	0.9862	0.2595	3.8
	3‰	500	1m46s	-	122.78	0s62	-	-	0.9916	0.1983	5.0
	4‰	560	2m1s	-	122.46	-	-	-	0.9852	0.1759	5.6
	5‰	510	2m10s	-	122.45	-	-	-	0.9850	0.1931	5.1
	6‰	580	2m8s	-	122.31	-	-	-	0.9822	0.1693	5.8
	7‰	660	2m23s	-	122.96	-	-	-	0.9952	0.1508	6.6
	8‰	600	2m18s	-	122.61	-	-	-	0.9882	0.1647	6.0
	9‰	650	2m19s	-	122.48	-	-	-	0.9856	0.1516	6.5
	10‰	600	2m32s	-	122.77	-	-	-	0.9914	0.1652	6.0
K_{12}	1‰	320	1m09s	-	122.55	-	-	-	0.9870	0.3084	3.2
	2‰	390	1m48s	-	123.03	-	-	-	0.9966	0.2555	3.9
	3‰	330	1m18s	-	122.09	-	-	-	0.9778	0.2963	3.3
	4‰	360	1m35s	-	123.06	-	-	-	0.9972	0.2770	3.6
	5‰	610	4m40s	-	122.98	-	-	-	0.9956	0.1632	6.1
	6‰	600	4m35s	-	122.29	-	-	-	0.9818	0.1636	6.0
	7‰	660	4m21s	-	122.51	-	-	-	0.9862	0.1494	6.6
	8‰	620	4m17s	-	122.21	-	-	-	0.9802	0.1581	6.2
	9‰	600	4m26s	-	121.95	-	-	-	0.9750	0.1625	6.0
	10‰	720	5m05s	1m06s	122.42	-	-	-	0.9844	0.1367	7.2

Appendix A. 3: Foaming test Nr. 3

Agents	Content	Measurements							Calculations		
		V_F	$t_{1/2}$	t_s	m_{50}	t_l	t_f	t_e	ρ_l	ρ_f	$V_G:V_L$
ABS	1‰	200	-	-	122.50	0s62	-	-	0.9860	0.4930	2.0
	2‰	400	1m10s	-	122.52	0s75	-	-	0.9864	0.2466	4.0
	3‰	490	1m41s	-	122.44	0s72	-	-	0.9848	0.2010	4.9
	4‰	550	2m22s	-	122.59	-	-	-	0.9878	0.1796	5.5
	5‰	650	2m14s	-	122.02	-	-	-	0.9764	0.1502	6.5
	6‰	660	2m12s	-	122.48	-	-	-	0.9856	0.1493	6.6
	7‰	660	2m8s	-	122.19	-	-	-	0.9798	0.1485	6.6
	8‰	550	2m13s	-	122.17	-	-	-	0.9794	0.1781	5.5
	9‰	550	2m01s	-	122.45	-	-	-	0.9850	0.1791	5.5
	10‰	570	1m45s	-	122.46	-	-	-	0.9852	0.1729	5.7
K_{12}	1‰	300	1m11s	-	122.09	-	-	-	0.9778	0.3259	3.0
	2‰	380	2m01s	-	123.03	-	-	-	0.9966	0.2623	3.8
	3‰	320	1m36s	-	122.46	-	-	-	0.9852	0.3079	3.2
	4‰	360	-	-	123.44	-	-	-	1.0048	0.2791	3.6
	5‰	580	4m25s	-	122.26	-	-	-	0.9812	0.1692	5.8
	6‰	620	4m28s	-	122.31	-	-	-	0.9822	0.1584	6.2
	7‰	610	4m30s	-	122.69	-	-	-	0.9898	0.1623	6.1
	8‰	800	5m41s	2m30s	122.97	-	-	-	0.9954	0.1245	8.0
	9‰	610	4m29s	-	123.28	-	-	-	1.0016	0.1642	6.1
	10‰	600	4m34s	40s	-	-	-	-	-	-	6.0

Appendix B

Appendix B. 1: HV-CMC Nr. 1

Agents	Content	Measurements							Calculations		
		V_F	$t_{1/2}$	t_s	m_{50}	t_l	t_f	t_e	ρ_l	ρ_f	$V_G : V_L$
4‰ ABS	1‰	550	3m04s	-	122.00	0s66	-	-	0.9760	0.1775	5.5
	2‰	430	6m30s	-	123.22	0s78	-	2s	1.0004	0.2327	4.3
	3‰	420	5m36s	51s	123.52	0s66	-	-	1.0064	0.2396	4.2
	4‰	420	8m31s	19s	121.07	0s78	-	-	0.9574	0.2280	4.2
	5‰	290	13m28s	-	123.49	0s96	-	-	1.0058	0.3469	2.9
	6‰	290	17m11s	-	122.40	1s37	-	-	0.9840	0.3393	2.9
	7‰	290	28m18s	-	121.74	1s44	-	-	0.9708	0.3348	2.9
	8‰	-	-	-	-	-	-	-	-	-	-
	9‰	-	-	-	-	-	-	-	-	-	-
	10‰	-	-	-	-	-	-	-	-	-	-
5‰ K$_{12}$	1‰	730	8m53s	-	120.65	0s85	-	10s	0.9490	0.1300	7.3
	2‰	600	10m45s	-	123.25	0s85	-	8s	1.0010	0.1668	6.0
	3‰	500	19m14s	-	121.80	0s94	-	11s	0.9720	0.1944	5.0
	4‰	490	23m59s	-	122.80	1s	-	15s	0.9920	0.2024	4.9
	5‰	480	29m15s	3m37s	122.35	0s78	-	30s	0.9830	0.2048	4.8
	6‰	500	42m46s	-	121.46	1s22	-	3s/18s	0.9652	0.1930	5.0
	7‰	500	-	-	122.28	1s09	-	2s/17s	0.9816	0.1963	5.0
	8‰	-	-	-	-	-	-	-	-	-	-
	9‰	-	-	-	-	-	-	-	-	-	-
	10‰	-	-	-	-	-	-	-	-	-	-

Appendix B. 2: HV-CMC Nr. 2

Agents	Content	Measurements							Calculations		
		V_F	$t_{1/2}$	t_s	m_{50}	t_l	t_f	t_e	ρ_l	ρ_f	$V_G : V_L$
	1‰	500	2m59s	-	121.95	0s84	-	-	0.9750	0.1950	5.0
	2‰	480	3m36s	-	121.80	0s63	-	-	0.9720	0.2025	4.8
	3‰	520	6m07s	20s	121.23	0s75	-	-	0.9606	0.1847	5.2
	4‰	440	10m04s	25s	122.37	0s96	-	-	0.9834	0.2235	4.4
4‰	5‰	360	13m34s	1m17s	121.48	0s91	-	-	0.9656	0.2682	3.6
ABS	6‰	320	16m32s	-	123.88	1s31	-	-	1.0136	0.3168	3.2
	7‰	270	20m46s	-	121.82	1s44	-	-	0.9724	0.3601	2.7
	8‰	-	-	-	-	-	-	-	-	-	-
	9‰	-	-	-	-	-	-	-	-	-	-
	10‰	-	-	-	-	-	-	-	-	-	-
	1‰	560	6m59s	-	123.87	0s91	-	10s	1.0134	0.1810	5.6
	2‰	580	10m46s	-	122.05	0s94	-	5s	0.9770	0.1684	5.8
	3‰	520	17m00s	-	123.90	1s03	-	8s	1.0140	0.1950	5.2
	4‰	520	23m18s	-	121.52	0s81	-	16s	0.9664	0.1859	5.2
5‰	5‰	450	27m24s	2m25s	121.50	1s13	-	30s	0.9660	0.2147	4.5
K_{12}	6‰	520	45m13s	4m22s	121.80	1s21	-	2s/18s	0.9720	0.1869	5.2
	7‰	550	-	-	120.61	1s29	-	2s/17s	0.9482	0.1724	5.5
	8‰	-	-	-	-	-	-	-	-	-	-
	9‰	-	-	-	-	-	-	-	-	-	-
	10‰	-	-	-	-	-	-	-	-	-	-

Appendix B. 3: HV-CMC Nr. 3

Agents	Content	Measurements							Calculations		
		V_F	$t_{1/2}$	t_s	m_{50}	t_l	t_f	t_e	ρ_l	ρ_f	$V_G : V_L$
4‰ ABS	1‰	520	3m22s	-	122.22	0s81	-	-	0.9804	0.1885	5.2
	2‰	500	2m41s	-	121.06	0s94	0s69	-	0.9572	0.1914	5.0
	3‰	480	5m25s	20s	122.20	0s81	0s88	-	0.9800	0.2041	4.8
	4‰	380	9m16s	21s	123.01	0s84	-	-	0.9962	0.2621	3.8
	5‰	320	14m20s	1m20s	122.29	1s06	-	-	0.9818	0.3068	3.2
	6‰	320	17m51s	-	121.40	-	-	-	0.9640	0.3013	3.2
	7‰	270	24m33s	-	123.52	1s52	-	-	1.0064	0.3727	2.7
	8‰	-	-	-	-	-	-	-	-	-	-
	9‰	-	-	-	-	-	-	-	-	-	-
	10‰	-	-	-	-	-	-	-	-	-	-
5‰ K_{12}	1‰	670	7m19s	-	121.08	0s72	22s09	8s	0.9576	0.1429	6.7
	2‰	570	10m58s	-	120.82	0s93	4s16	8s	0.9524	0.1671	5.7
	3‰	490	18m08s	-	122.48	0s85	37s97	8s	0.9856	0.2011	4.9
	4‰	510	20m36s	-	123.54	0s97	45s38	12s	1.0068	0.1974	5.1
	5‰	430	34m16s	5m13s	123.53	1s21	1m16s	28s	1.0066	0.2341	4.3
	6‰	500	>60min	-	124.14	1s37	1m14s	2s/18s	1.0188	0.2038	5.0
	7‰	530	>60min	-	121.37	1s43	1m40s	2s/18s	0.9634	0.1818	5.3
	8‰	-	-	-	-	-	-	-	-	-	-
	9‰	-	-	-	-	-	-	-	-	-	-
	10‰	-	-	-	-	-	-	-	-	-	-

Appendix B. 4: HEC Nr. 1

Agents	Contents	Measurements							Calculations		
		V_F	$t_{1/2}$	t_s	m_{50}	t_l	t_f	t_e	ρ_l	ρ_f	$V_G : V_L$
4‰ ABS	1‰	500	5m17s	-	121.84	0s82	-	-	0.9728	0.1946	5.0
	2‰	400	11m46s	-	123.46	0s88	-	-	1.0052	0.2513	4.0
	3‰	300	17m39s	-	120.96	1s63	-	-	0.9552	0.3184	3.0
	4‰	-	-	-	-	-	-	-	-	-	-
	5‰	-	-	-	-	-	-	-	-	-	-
	6‰	-	-	-	-	-	-	-	-	-	-
	7‰	-	-	-	-	-	-	-	-	-	-
	8‰	-	-	-	-	-	-	-	-	-	-
	9‰	-	-	-	-	-	-	-	-	-	-
	10‰	-	-	-	-	-	-	-	-	-	-
5‰ K_{12}	1‰	570	12m26s	-	122.23	0s75	-	15s	0.9806	0.1720	5.7
	2‰	430	23m01s	-	121.13	1s06	-	13s	0.9586	0.2229	4.3
	3‰	440	36m05s	-	122.75	1s21	-	20s	0.9910	0.2252	4.4
	4‰	380	47m13s	-	123.13	2s57	-	36s	0.9986	0.2628	3.8
	5‰	330	>60min	7m49s	121.19	4s47	-	50s	0.9598	0.2909	3.3
	6‰	300	>60min	13m22s	122.01	9s60	-	-	0.9762	0.3254	3.0
	7‰	300	>60min	17m13s	121.97	22s63	-	2s	0.9754	0.3251	3.0
	8‰	300	>60min	-	122.02	37s53	-	2s	0.9764	0.3255	3.0
	9‰	-	-	-	-	-	-	-	-	-	-
	10‰	-	-	-	-	-	-	-	-	-	-

Appendix B. 5: HEC Nr. 2

Agents	Content	Measurements							Calculations		
		V_F	$t_{1/2}$	t_s	m_{50}	t_l	t_f	t_e	ρ_l	ρ_f	$V_G : V_L$
4‰ ABS	1‰	480	4m56s	-	122.04	0s81	-	-	0.9768	0.2035	4.8
	2‰	380	10m11s	11s	122.78	1s09	-	-	0.9916	0.2610	3.8
	3‰	280	18m51s	-	123.00	1s47	-	-	0.9960	0.3557	2.8
	4‰	-	-	-	-	-	-	-	-	-	-
	5‰	-	-	-	-	-	-	-	-	-	-
	6‰	-	-	-	-	-	-	-	-	-	-
	7‰	-	-	-	-	-	-	-	-	-	-
	8‰	-	-	-	-	-	-	-	-	-	-
	9‰	-	-	-	-	-	-	-	-	-	-
	10‰	-	-	-	-	-	-	-	-	-	-
5‰ K_{12}	1‰	580	11m44s	-	123.70	0s90	-	15s	1.0100	0.1741	5.8
	2‰	500	22m57s	-	123.95	1s19	-	13s	1.0150	0.2030	5.0
	3‰	420	32m53s	-	121.52	1s44	-	26s	0.9664	0.2301	4.2
	4‰	360	49m18s	-	122.72	2s63	-	20s	0.9904	0.2751	3.6
	5‰	220	>60min	6m48s	122.22	5s69	-	-	0.9804	0.4456	2.2
	6‰	300	>60min	13m36s	120.72	8s13	-	2s	0.9504	0.3168	3.0
	7‰	300	>60min	16m06s	123.21	14s16	-	2s	1.0002	0.3334	3.0
	8‰	300	>60min	-	120.34	33s15	-	2s	0.9428	0.3143	3.0
	9‰	-	-	-	-	-	-	-	-	-	-
	10‰	-	-	-	-	-	-	-	-	-	-

Appendix B. 6: HEC Nr. 3

Agents	Content	Measurements							Calculations		
		V_F	$t_{1/2}$	t_s	m_{50}	t_l	t_f	t_e	ρ_l	ρ_f	$V_G:V_L$
4‰ ABS	1‰	520	5m37s	-	121.17	1s10	-	-	0.9594	0.1845	5.2
	2‰	400	10m19s	22s	121.87	0s85	-	-	0.9734	0.2434	4.0
	3‰	290	17m31s	-	122.70	1s63	-	-	0.9900	0.3414	2.9
	4‰	-	-	-	-	-	-	-	-	-	-
	5‰	-	-	-	-	-	-	-	-	-	-
	6‰	-	-	-	-	-	-	-	-	-	-
	7‰	-	-	-	-	-	-	-	-	-	-
	8‰	-	-	-	-	-	-	-	-	-	-
	9‰	-	-	-	-	-	-	-	-	-	-
	10‰	-	-	-	-	-	-	-	-	-	-
5‰ K_{12}	1‰	560	11m54s	-	121.34	0s93	39s06	15s	0.9628	0.1719	5.6
	2‰	490	22m55s	-	123.41	1s16	1m13s	14s	1.0042	0.2049	4.9
	3‰	400	36m08s	-	123.21	1s50	1m11s	24s	1.0002	0.2501	4.0
	4‰	390	49m54s	-	121.35	2s60	1m54s	21s	0.9630	0.2469	3.9
	5‰	360	>60min	11m11s	122.90	4s40	27s16	20s	0.9940	0.2761	3.6
	6‰	350	>60min	11m32s	121.92	8s78	-	2s	0.9744	0.2784	3.5
	7‰	350	>60min	17m41s	121.19	15s81	-	2s	0.9598	0.2742	3.5
	8‰	300	-	-	120.14	37s09	-	2s	0.9388	0.3129	3.0
	9‰	-	-	-	-	-	-	-	-	-	-
	10‰	-	-	-	-	-	-	-	-	-	-

Appendix B. 7: LV-PAC Nr. 1

Agents	Content	Measurements							Calculations		
		V_F	$t_{1/2}$	t_s	m_{50}	t_l	t_f	t_e	ρ_l	ρ_f	$V_G:V_L$
4‰ ABS	1‰	480	2m31s	-	122.31	0s79	-	-	0.9822	0.2046	4.8
	2‰	500	2m12s	-	122.74	0s62	-	-	0.9908	0.1982	5.0
	3‰	500	3m37s	-	121.25	0s88	-	-	0.9610	0.1922	5.0
	4‰	400	4m21s	-	122.31	0s82	-	-	0.9822	0.2456	4.0
	5‰	380	2m49s	-	123.40	0s81	-	-	1.0040	0.2642	3.8
	6‰	340	2m43s	-	122.19	0s69	-	-	0.9798	0.2882	3.4
	7‰	370	3m21s	-	121.96	0s87	-	-	0.9752	0.2636	3.7
	8‰	360	4m11s	-	123.60	0s78	-	-	1.0080	0.2800	3.6
	9‰	310	?	-	123.59	0s82	-	-	1.0078	0.3251	3.1
	10‰	290	2m54s	-	122.02	0s84	-	-	0.9764	0.3367	2.9
5‰ K$_{12}$	1‰	610	5m32s	-	123.15	0s84	-	<1s	0.9990	0.1638	6.1
	2‰	600	7m08s	-	122.34	0s87	-	5s	0.9828	0.1638	6.0
	3‰	610	7m36s	-	121.33	0s81	-	9s	0.9626	0.1578	6.1
	4‰	570	8m19s	-	121.94	0s94	-	6s	0.9748	0.1710	5.7
	5‰	510	9m33s	-	122.73	0s75	-	2s	0.9906	0.1942	5.1
	6‰	530	-	-	122.06	0s84	-	2s	0.9772	0.1844	5.3
	7‰	520	12m24s	-	123.68	0s75	-	1s/30s	1.0096	0.1942	5.2
	8‰	510	13m07s	-	121.96	0s81	-	1s/20s	0.9752	0.1912	5.1
	9‰	520	14m37s	2m19s	123.87	1s06	-	1s/26s	1.0134	0.1949	5.2
	10‰	490	15m59s	-	122.10	0s94	-	1s/30s	0.9780	0.1996	4.9

Appendix B. 8: LV-PAC Nr. 2

Agents	Content	Measurements							Calculations		
		V_F	$t_{1/2}$	t_s	m_{50}	t_l	t_f	t_e	ρ_l	ρ_f	$V_G : V_L$
4‰ ABS	1‰	480	2m32s	-	122.78	0s72	-	-	0.9916	0.2066	4.8
	2‰	510	2m57s	-	121.18	0s75	-	-	0.9596	0.1881	5.1
	3‰	520	3m21s	-	123.77	0s72	-	-	1.0114	0.1945	5.2
	4‰	460	2m37s	-	121.43	0s90	-	-	0.9646	0.2097	4.6
	5‰	380	2m33s	-	122.30	1s09	-	-	0.9820	0.2584	3.8
	6‰	370	3m49s	-	121.28	0s88	-	-	0.9616	0.2599	3.7
	7‰	380	3m15s	-	124.32	0s97	-	-	1.0224	0.2691	3.8
	8‰	350	4m09s	-	121.03	0s88	-	-	0.9566	0.2733	3.5
	9‰	320	2m32s	-	121.01	0s81	-	-	0.9562	0.2988	3.2
	10‰	320	3m38s	-	121.46	0s88	-	-	0.9652	0.3016	3.2
5‰ K_{12}	1‰	610	5m55s	-	121.33	1s	-	<1s	0.9626	0.1578	6.1
	2‰	610	6m51s	-	121.65	0s81	-	5s	0.9690	0.1589	6.1
	3‰	600	8m12s	-	123.17	0s87	-	3s	0.9994	0.1666	6.0
	4‰	540	8m06s	-	123.30	0s81	-	2s	1.0020	0.1856	5.4
	5‰	570	9m15s	-	121.80	0s85	-	2s	0.9720	0.1706	5.7
	6‰	580	10m21s	-	122.98	0s93	-	2s	0.9956	0.1717	5.8
	7‰	530	13m01s	-	121.47	1s	-	1s/24s	0.9654	0.1822	5.3
	8‰	490	13m21s	2m09s	123.41	0s97	-	2s/20s	1.0042	0.2049	4.9
	9‰	500	15m20s	-	121.63	0s78	-	2s/30s	0.9686	0.1938	5.0
	10‰	490	15m48s	-	122.93	0s82	-	2s/31s	0.9946	0.2030	4.9

Appendix B. 9: LV-PAC Nr. 3

Agents	Content	Measurements							Calculations		
		V_F	$t_{1/2}$	t_s	m_{50}	t_l	t_f	t_e	ρ_l	ρ_f	$V_G : V_L$
4‰ ABS	1‰	520	3m05s	-	123.02	0s69	-	-	0.9964	0.1916	5.2
	2‰	550	3m10s	-	121.90	0s78	0s68	-	0.9740	0.1771	5.5
	3‰	460	3m17s	-	122.62	0s75	0s69	-	0.9884	0.2149	4.6
	4‰	430	2m39s	-	123.67	0s72	-	-	1.0094	0.2347	4.3
	5‰	380	2m32s	-	121.42	0s90	-	-	0.9644	0.2538	3.8
	6‰	370	3m19s	-	123.74	0s84	-	-	1.0108	0.2732	3.7
	7‰	400	4m12s	-	121.81	0s85	-	-	0.9722	0.2431	4.0
	8‰	370	4m07s	-	122.14	0s81	-	-	0.9788	0.2645	3.7
	9‰	370	3m14s	-	122.65	0s97	-	-	0.9890	0.2673	3.7
	10‰	330	3m14s	-	123.55	0s81	-	-	1.0070	0.3052	3.3
5‰ K$_{12}$	1‰	570	5m36s	-	121.87	0s78	25s22	<1s	0.9734	0.1708	5.7
	2‰	610	6m56s	-	124.17	0s91	16s60	4s39	1.0194	0.1671	6.1
	3‰	560	7m53s	-	121.75	0s81	20s59	3s	0.9710	0.1734	5.6
	4‰	580	8m11s	-	121.51	0s84	5s06	2s	0.9662	0.1666	5.8
	5‰	540	9m58s	-	121.29	0s81	20s91	2s	0.9618	0.1781	5.4
	6‰	560	8m31s	-	121.34	0s78	26s22	2s	0.9628	0.1713	5.6
	7‰	550	12m45s	-	122.20	0s94	22s66	2s/25s	0.9800	0.1782	5.5
	8‰	510	13m05s	2m17s	122.42	0s97	23s50	2s/20s	0.9844	0.1930	5.1
	9‰	520	15m09s	1m43s	123.71	1s06	21s13	1s/24s	1.0102	0.1943	5.2
	10‰	500	14m52s	-	124.24	0s82	21s25	2s/26s	1.0208	0.2041	5.0

Appendix B. 10: Sesbania Nr. 1

Agents	Content	Measurements							Calculations		
		V_F	$t_{1/2}$	t_s	m_{50}	t_1	t_f	t_e	ρ_l	ρ_f	$V_G : V_L$
4‰ ABS	1‰	600	3m58s	-	122.38	0s60	-	-	0.9836	0.1639	6.0
	2‰	550	13m51s	-	121.41	0s75	-	1s59	0.9642	0.1753	5.5
	3‰	480	31m52s	23s	122.06	1s16	-	<1s	0.9772	0.2036	4.8
	4‰	430	>60min	6m11s	121.19	2s09	-	<2s	0.9598	0.2232	4.3
	5‰	420	>60min	1m19s	121.60	9s44	-	3~4s	0.9680	0.2305	4.2
	6‰	350	>60min	-	123.03	27s	-	-	0.9966	0.2847	3.5
	7‰	290	>60min	3m48s	121.35	57s47	-	3~4s	0.9630	0.3321	2.9
	8‰	-	-	-	-	-	-	-	-	-	-
	9‰	-	-	-	-	-	-	-	-	-	-
	10‰	-	-	-	-	-	-	-	-	-	-
5‰ K$_{12}$	1‰	640	7m52s	-	121.98	0s71	-	3s48	0.9756	0.1524	6.4
	2‰	960	29m02s	3m53s	122.29	1s	-	2s/30s	0.9818	0.1023	9.6
	3‰	900	>60min	9m53s	123.45	1s13	-	2s/17s	1.0050	0.1117	9.0
	4‰	880	>60min	12m46s	121.21	2s43	-	1s/7s	0.9602	0.1091	8.8
	5‰	850	>60min	25m47s	123.38	7s68	-	2s/7s	1.0036	0.1181	8.5
	6‰	490	>60min	26m34s	122.52	23s31	-	4s	0.9864	0.2013	4.9
	7‰	560	>60min	38m38s	122.62	56s75	-	2s/5s	0.9884	0.1765	5.6
	8‰	450	-	-	122.60	1m23s	-	2s/5s	0.9880	0.2196	4.5
	9‰	-	-	-	-	-	-	-	-	-	-
	10‰	-	-	-	-	-	-	-	-	-	-

Appendix B. 11: Sesbania Nr. 2

Agents	Content	Measurements							Calculations		
		V_F	$t_{1/2}$	t_s	m_{50}	t_l	t_f	t_e	ρ_l	ρ_f	$V_G : V_L$
4‰ ABS	1‰	600	3m52s	-	123.63	0s69	-	-	1.0086	0.1681	6.0
	2‰	600	13m56s	-	122.23	0s94	-	2s	0.9806	0.1634	6.0
	3‰	480	33m11s	1m23s	121.19	1s37	-	<1s	0.9598	0.2000	4.8
	4‰	430	>60min	5m38s	122.31	2s88	-	<2s	0.9822	0.2284	4.3
	5‰	390	>60min	1m36s	121.01	10s94	-	3s	0.9562	0.2452	3.9
	6‰	420	>60min	-	120.56	30s69	-	2s	0.9472	0.2255	4.2
	7‰	310	>60min	3m39s	119.93	53s90	-	4s	0.9346	0.3015	3.1
	8‰	-	-	-	-	-	-	-	-	-	-
	9‰	-	-	-	-	-	-	-	-	-	-
	10‰	-	-	-	-	-	-	-	-	-	-
5‰ K_{12}	1‰	640	8m13s	-	123.33	0s84	-	4s	1.0026	0.1567	6.4
	2‰	970	29m23s	4m05s	121.16	0s84	-	2s/26s	0.9592	0.0989	9.7
	3‰	820	>60min	9m31s	122.20	1s16	-	2s/14s	0.9800	0.1195	8.2
	4‰	850	>60min	11m43s	120.82	2s25	-	1s/14s	0.9524	0.1120	8.5
	5‰	900	>60m	20m23s	121.42	8s63	-	1s/5s	0.9644	0.1072	9.0
	6‰	920	>60min	28m47s	120.89	27s38	-	3s/7s	0.9538	0.1037	9.2
	7‰	530	>60min	29m09s	122.82	55s25	-	2s/4s	0.9924	0.1872	5.3
	8‰	400	>60min	-	123.62	1m33s	-	2s/4s	1.0084	0.2521	4.0
	9‰	-	-	-	-	-	-	-	-	-	-
	10‰	-	-	-	-	-	-	-	-	-	-

Appendix B. 12: Sesbania Nr. 3

Agents	Content	Measurements							Calculations		
		V_F	$t_{1/2}$	t_s	m_{50}	t_i	t_f	t_e	ρ_i	ρ_f	$V_G:V_L$
4‰ ABS	1‰	600	-	-	120.76	0s75	0s63	-	0.9512	0.1585	6.0
	2‰	550	14m09s	-	123.11	0s91	3s91	2s	0.9982	0.1815	5.5
	3‰	480	33m39s	-	122.89	1s	1m	-	0.9938	0.2070	4.8
	4‰	440	>60m	5m53s	123.80	3s	2m11s	<2s	1.0120	0.2300	4.4
	5‰	440	>60m	1m13s	123.51	11s69	3m26s	3s	1.0062	0.2287	4.4
	6‰	350	>60m	23m33s	122.00	30s40	5m03s	3s	0.9760	0.2789	3.5
	7‰	320	>60m	3m34s	119.29	1m06s	6m15s	4s	0.9218	0.2881	3.2
	8‰	-	-	-	-	-	-	-	-	-	-
	9‰	-	-	-	-	-	-	-	-	-	-
	10‰	-	-	-	-	-	-	-	-	-	-
5‰ K$_{12}$	1‰	580	7m33s	-	121.48	0s81	31s40	3s81	0.9656	0.1665	5.8
	2‰	900	29m41s	4m29s	123.45	0s94	40s03	2s/21s	1.0050	0.1117	9.0
	3‰	890	>60min	12m09s	121.17	1s09	1m11s	1s/7s	0.9594	0.1078	8.9
	4‰	800	>60min	-	122.71	2s88	2m14s	1s/12s	0.9902	0.1238	8.0
	5‰	830	>60min	34m55s	120.62	8s37	3m17s	1s/14s	0.9484	0.1143	8.3
	6‰	850	>60min	32m29s	120.52	28s25	4m52s	3s/8s	0.9464	0.1113	8.5
	7‰	520	>60min	28m00s	120.86	1m11s	6m59s	2s/4s	0.9532	0.1833	5.2
	8‰	480	>60min	-	121.49	1m20s	-	2s/5s	0.9658	0.2012	4.8
	9‰	-	-	-	-	-	-	-	-	-	-
	10‰	-	-	-	-	-	-	-	-	-	-

Appendix B. 13: HV-PAC Nr. 1

Agents	Content	Measurements							Calculations		
		V_F	$t_{1/2}$	t_s	m_{50}	t_l	t_f	t_e	ρ_l	ρ_f	$V_G : V_L$
	1‰	500	4m33s	-	121.46	0s63	-	-	0.9652	0.1930	5.0
	2‰	480	7m12s	-	123.43	0s94	-	-	1.0046	0.2093	4.8
	3‰	310	8m41s	-	123.34	1s28	-	-	1.0028	0.3235	3.1
	4‰	270	11m49s	-	123.36	1s09	-	4s	1.0032	0.3716	2.7
4‰	5‰	260	16m11s	51s	123.58	1s94	-	-	1.0076	0.3875	2.6
ABS	6‰	240	16m6s	-	123.61	1s72	-	-	1.0082	0.4201	2.4
	7‰	-	-	-	-	-	-	-	-	-	-
	8‰	-	-	-	-	-	-	-	-	-	-
	9‰	-	-	-	-	-	-	-	-	-	-
	10‰	-	-	-	-	-	-	-	-	-	-
	1‰	580	8m52s	-	122.86	0s68	-	<1s	0.9932	0.1712	5.8
	2‰	550	14m39s	-	121.49	0s81	-	9s	0.9658	0.1756	5.5
	3‰	450	21m18s	-	121.79	0s97	-	10s	0.9718	0.2160	4.5
	4‰	440	33m26s	-	122.90	1s06	-	18s	0.9940	0.2259	4.4
5‰	5‰	280	33m19s	-	122.40	1s41	-	4s	0.9840	0.3514	2.8
K_{12}	6‰	300	47m12s	-	121.45	2s18	-	16s	0.9650	0.3217	3.0
	7‰	320	55m10s	3m19s	121.37	3s75	-	2s	0.9634	0.3011	3.2
	8‰	340	-	-	123.99	3s84	-	2s	1.0158	0.2988	3.4
	9‰	-	-	-	-	-	-	-	-	-	-
	10‰	-	-	-	-	-	-	-	-	-	-

Appendix B. 14: HV-PAC Nr. 2

Agents	Content	Measurements							Calculations		
		V_F	$t_{1/2}$	t_s	m_{50}	t_l	t_f	t_e	ρ_l	ρ_f	$V_G:V_L$
4‰ ABS	1‰	510	4m28s	-	121.94	0s78	-	-	0.9748	0.1911	5.1
	2‰	460	6m30s	-	121.29	0s87	-	-	0.9618	0.2091	4.6
	3‰	360	9m10s	-	121.50	1s	-	-	0.9660	0.2683	3.6
	4‰	300	15m52s	-	121.45	1s19	-	4s	0.9650	0.3217	3.0
	5‰	270	16m58s	49s	122.57	1s35	-	-	0.9874	0.3657	2.7
	6‰	240	19m56s	-	121.15	2s16	-	-	0.9590	0.3996	2.4
	7‰	-	-	-	-	-	-	-	-	-	-
	8‰	-	-	-	-	-	-	-	-	-	-
	9‰	-	-	-	-	-	-	-	-	-	-
	10‰	-	-	-	-	-	-	-	-	-	-
5‰ K_{12}	1‰	570	9m57s	-	122.03	0s81	-	<1s	0.9766	0.1713	5.7
	2‰	530	15m58s	-	122.12	0s91	-	-	0.9784	0.1846	5.3
	3‰	460	21m22s	-	122.91	1s21	-	8s	0.9942	0.2161	4.6
	4‰	370	25m52s	-	121.59	1s09	-	-	0.9678	0.2616	3.7
	5‰	400	35m00s	-	121.60	1s28	-	4s	0.9680	0.2420	4.0
	6‰	270	41m33s	-	122.86	3s03	-	-	0.9932	0.3679	2.7
	7‰	400	58m02s	8m40s	121.15	3s72	-	2s/58s	0.9590	0.2398	4.0
	8‰	350	-	-	120.97	3s37	-	2s	0.9554	0.2730	3.5
	9‰	-	-	-	-	-	-	-	-	-	-
	10‰	-	-	-	-	-	-	-	-	-	-

Appendix B. 15: HV-PAC Nr. 3

Agents	Content	Measurements							Calculations		
		V_F	$t_{1/2}$	t_s	m_{50}	t_l	t_f	t_e	ρ_l	ρ_f	$V_G : V_L$
4‰ ABS	1‰	490	4m33s	-	123.45	0s72	-	-	1.0050	0.2051	4.9
	2‰	450	6m14s	-	121.72	0s81	-	-	0.9704	0.2156	4.5
	3‰	350	10m43s	-	122.58	1s31	-	-	0.9876	0.2822	3.5
	4‰	300	14m58s	-	121.85	1s31	-	5s	0.9730	0.3243	3.0
	5‰	280	17m45s	-	123.62	1s44	-	-	1.0084	0.3601	2.8
	6‰	240	23m39s	-	123.20	2s31	-	-	1.0000	0.4167	2.4
	7‰	-	-	-	-	-	-	-	-	-	-
	8‰	-	-	-	-	-	-	-	-	-	-
	9‰	-	-	-	-	-	-	-	-	-	-
	10‰	-	-	-	-	-	-	-	-	-	-
5‰ K_{12}	1‰	610	9m34s	-	120.99	0s88	47s50	1s	0.9558	0.1567	6.1
	2‰	520	15m28s	-	123.44	0s93	49s03	8~9s	1.0048	0.1932	5.2
	3‰	430	21m53s	-	121.47	1s34	1m11s	10s	0.9654	0.2245	4.3
	4‰	450	28m42s	-	121.42	0s97	1m09s	15s	0.9644	0.2143	4.5
	5‰	420	34m56s	-	123.53	1s40	16s50	3s	1.0066	0.2397	4.2
	6‰	390	40m23s	-	120.32	2s19	25s85	9s	0.9424	0.2416	3.9
	7‰	370	>60min	8m21s	121.45	3s22	-	2s	0.9650	0.2608	3.7
	8‰	350	>60min	-	121.45	3s50	-	2s	0.9650	0.2757	3.5
	9‰	-	-	-	-	-	-	-	-	-	-
	10‰	-	-	-	-	-	-	-	-	-	-

Appendix B. 16: Konjac Nr. 1

Agents	Content	Measurements							Calculations		
		V_F	$t_{1/2}$	t_s	m_{50}	t_l	t_f	t_e	ρ_l	ρ_f	$V_G : V_L$
4‰ ABS	1‰	550	2m30s	-	124.02	0s69	-	-	1.0164	0.1848	5.5
	2‰	550	12m47s	-	121.90	0s94	-	-	0.9740	0.1771	5.5
	3‰	390	35m58s	-	123.94	1s47	-	-	1.0148	0.2602	3.9
	4‰	260	>60min	6m04s	121.41	7s03	-	2s	0.9642	0.3708	2.6
	5‰	-	-	-	-	-	-	-	-	-	-
	6‰	-	-	-	-	-	-	-	-	-	-
	7‰	-	-	-	-	-	-	-	-	-	-
	8‰	-	-	-	-	-	-	-	-	-	-
	9‰	-	-	-	-	-	-	-	-	-	-
	10‰	-	-	-	-	-	-	-	-	-	-
5‰ K_{12}	1‰	610	6m33s	-	121.61	0s81	-	1s28	0.9682	0.1587	6.1
	2‰	890	15m54s	-	121.44	0s84	-	3s/30s	0.9648	0.1084	8.9
	3‰	800	60m21s	10m30s	123.89	2s50	-	2s/29	1.0138	0.1267	8.0
	4‰	420	>60min	6m49s	121.52	8s59	-	2s	0.9664	0.2301	4.2
	5‰	360	>60min	13m04s	121.68	21s19	-	4s	0.9696	0.2693	3.6
	6‰	520	>60min	-	120.78	42s	-	2s/5s	0.9516	0.1830	5.2
	7‰	340	>60min	-	122.09	2m08s	-	2s	0.9778	0.2876	3.4
	8‰	-	-	-	-	-	-	-	-	-	-
	9‰	-	-	-	-	-	-	-	-	-	-
	10‰	-	-	-	-	-	-	-	-	-	-

Appendix B. 17: Konjac Nr. 2

Agents	Content	Measurements							Calculation		
		V_F	$t_{1/2}$	t_s	m_{50}	t_l	t_f	t_e	ρ_l	ρ_f	$V_G : V_L$
4‰ ABS	1‰	580	2m50s	-	122.04	0s72	-	-	0.9768	0.1684	5.8
	2‰	560	11m11s	29s	123.73	0s84	-	-	1.0106	0.1805	5.6
	3‰	380	38m01s	-	121.63	1s63	-	-	0.9686	0.2549	3.8
	4‰	230	>60min	3m01s	123.06	7s50	-	2s	0.9972	0.4336	2.3
	5‰	-	-	-	-	-	-	-	-	-	-
	6‰	-	-	-	-	-	-	-	-	-	-
	7‰	-	-	-	-	-	-	-	-	-	-
	8‰	-	-	-	-	-	-	-	-	-	-
	9‰	-	-	-	-	-	-	-	-	-	-
	10‰	-	-	-	-	-	-	-	-	-	-
5‰ K_{12}	1‰	900	9m21s	-	123.72	0s97	-	7s	1.0104	0.1123	9.0
	2‰	560	10m05s	-	123.59	0s87	-	3s	1.0078	0.1710	5.6
	3‰	460	55m48s	6m03s	121.97	2s88	-	1s26	0.9754	0.2120	4.6
	4‰	360	>60min	2m15s	120.17	8s	-	2s	0.9394	0.2609	3.6
	5‰	340	>60min	14m29s	121.71	27s13	-	4s	0.9702	0.2854	3.4
	6‰	560	>60min	-	122.98	47s44	-	1s/3s	0.9956	0.1778	5.6
	7‰	390	>60min	-	121.13	2m04s	-	2s	0.9586	0.2458	3.9
	8‰	-	-	-	-	-	-	-	-	-	-
	9‰	-	-	-	-	-	-	-	-	-	-
	10‰	-	-	-	-	-	-	-	-	-	-

Appendix B. 18: Konjac Nr. 3

Agents	Content	Measurements							Calculation		
		V_F	$t_{1/2}$	t_s	m_{50}	t_l	t_f	t_e	ρ_l	ρ_f	$V_G : V_L$
4‰ ABS	1‰	500	2m59s	-	121.45	0s69	0s81	-	0.9650	0.1930	5.0
	2‰	530	12m11s	37s	121.06	0s78	1s41	-	0.9572	0.1806	5.3
	3‰	380	34m35s	-	122.11	1s72	-	-	0.9782	0.2574	3.8
	4‰	250	>60min	-	121.89	9s66	-	2s	0.9738	0.3895	2.5
	5‰	-	-	-	-	-	-	-	-	-	-
	6‰	-	-	-	-	-	-	-	-	-	-
	7‰	-	-	-	-	-	-	-	-	-	-
	8‰	-	-	-	-	-	-	-	-	-	-
	9‰	-	-	-	-	-	-	-	-	-	-
	10‰	-	-	-	-	-	-	-	-	-	-
5‰ K_{12}	1‰	630	6m19s	-	122.43	0s96	3s41	6s	0.9846	0.1563	6.3
	2‰	550	10m13s	-	121.90	0s94	31s47	3s	0.9740	0.1771	5.5
	3‰	450	52m33s	3m49s	120.94	3s	1m38s	1s23	0.9548	0.2122	4.5
	4‰	950	>60min	18m40s	122.55	9s57	2m10s	2s/5s	0.9870	0.1039	9.5
	5‰	410	>60min	15m05s	123.44	29s93	4m18s	2s	1.0048	0.2451	4.1
	6‰	520	>60min	-	123.14	1m02s	4m18s	2s/3s	0.9988	0.1921	5.2
	7‰	360	>60min	-	123.52	2m40s	-	-	1.0064	0.2796	3.6
	8‰	-	-	-	-	-	-	-	-	-	-
	9‰	-	-	-	-	-	-	-	-	-	-
	10‰	-	-	-	-	-	-	-	-	-	-

Appendix B. 19: XC Nr. 1

Agents	Content	Measurements							Calculations		
		V_F	$t_{1/2}$	t_s	m_{50}	t_l	t_f	t_e	ρ_l	ρ_f	$V_G:V_L$
4‰ ABS	1‰	450	16m53s	30s	121.34	0s75	-	-	0.9628	0.2140	4.5
	2‰	360	60m51s	-	121.53	1s32	-	-	0.9666	0.2685	3.6
	3‰	210	>60min	-	118.95	8s16	-	-	0.9150	0.4357	2.1
	4‰	-	-	-	-	-	-	-	-	-	-
	5‰	-	-	-	-	-	-	-	-	-	-
	6‰	-	-	-	-	-	-	-	-	-	-
	7‰	-	-	-	-	-	-	-	-	-	-
	8‰	-	-	-	-	-	-	-	-	-	-
	9‰	-	-	-	-	-	-	-	-	-	-
	10‰	-	-	-	-	-	-	-	-	-	-
5‰ K_{12}	1‰	500	27m39s	-	121.16	0s75	-	-	0.9592	0.1918	5.0
	2‰	450	>60min	8m32s	120.24	1s78	-	21s	0.9408	0.2091	4.5
	3‰	410	>60min	9m50s	120.85	6s10	-	16s	0.9530	0.2324	4.1
	4‰	220	>60min	>60m	118.68	37s56	-	-	0.9096	0.4135	2.2
	5‰	-	-	-	-	-	-	-	-	-	-
	6‰	-	-	-	-	-	-	-	-	-	-
	7‰	-	-	-	-	-	-	-	-	-	-
	8‰	-	-	-	-	-	-	-	-	-	-
	9‰	-	-	-	-	-	-	-	-	-	-
	10‰	-	-	-	-	-	-	-	-	-	-

Appendix B. 20: XC Nr. 2

Agents	Content	Measurements							Calculations		
		V_F	$t_{1/2}$	t_s	m_{50}	t_l	t_f	t_e	ρ_l	ρ_f	$V_G : V_L$
4‰ ABS	1‰	480	21m04s	2m20s	121.43	1s03	-	-	0.9646	0.2010	4.8
	2‰	300	56m16s	41s	121.10	1s35	-	-	0.9580	0.3193	3.0
	3‰	220	>60min	-	119.27	7s72	-	15s	0.9214	0.4188	2.2
	4‰	-	-	-	-	-	-	-	-	-	-
	5‰	-	-	-	-	-	-	-	-	-	-
	6‰	-	-	-	-	-	-	-	-	-	-
	7‰	-	-	-	-	-	-	-	-	-	-
	8‰	-	-	-	-	-	-	-	-	-	-
	9‰	-	-	-	-	-	-	-	-	-	-
	10‰	-	-	-	-	-	-	-	-	-	-
5‰ K_{12}	1‰	490	28m12s	-	121.99	0s78	-	-	0.9758	0.1991	4.9
	2‰	420	>60min	5m19s	121.37	1s75	-	-	0.9634	0.2294	4.2
	3‰	420	>60min	2m39s	120.91	7s06	-	3s	0.9542	0.2272	4.2
	4‰	250	>60min	-	116.97	40s25	-	-	0.8754	0.3502	2.5
	5‰	-	-	-	-	-	-	-	-	-	-
	6‰	-	-	-	-	-	-	-	-	-	-
	7‰	-	-	-	-	-	-	-	-	-	-
	8‰	-	-	-	-	-	-	-	-	-	-
	9‰	-	-	-	-	-	-	-	-	-	-
	10‰	-	-	-	-	-	-	-	-	-	-

Appendix B. 21: XC Nr. 3

Agents	Content	Measurements							Calculations		
		V_F	$t_{1/2}$	t_s	m_{50}	t_l	t_f	t_e	ρ_l	ρ_f	$V_G : V_L$
4‰ ABS	1‰	460	18m09s	2m30s	121.42	0s75	0s57	-	0.9644	0.2097	4.6
	2‰	320	>78min	27s	122.93	1s43	-	-	0.9946	0.3108	3.2
	3‰	220	>60min	-	119.91	9s03	-	-	0.9342	0.4246	2.2
	4‰	-	-	-	-	-	-	-	-	-	-
	5‰	-	-	-	-	-	-	-	-	-	-
	6‰	-	-	-	-	-	-	-	-	-	-
	7‰	-	-	-	-	-	-	-	-	-	-
	8‰	-	-	-	-	-	-	-	-	-	-
	9‰	-	-	-	-	-	-	-	-	-	-
	10‰	-	-	-	-	-	-	-	-	-	-
5‰ K_{12}	1‰	490	29m24s	-	123.27	0s78	7s13	-	1.0014	0.2044	4.9
	2‰	390	>60min	9m01s	123.89	2s	-	-	1.0138	0.2599	3.9
	3‰	390	>60min	18m54s	122.91	7s19	1m27s	18s	0.9942	0.2549	3.9
	4‰	330	>6min	>60min	118.60	37s56	2m55s	5s	0.9080	0.2752	3.3
	5‰	-	-	-	-	-	-	-	-	-	-
	6‰	-	-	-	-	-	-	-	-	-	-
	7‰	-	-	-	-	-	-	-	-	-	-
	8‰	-	-	-	-	-	-	-	-	-	-
	9‰	-	-	-	-	-	-	-	-	-	-
	10‰	-	-	-	-	-	-	-	-	-	-

Appendix C

Appendix C. 1: 5‰ K_{12} + 0.6‰ Sesbania Nr. 1

Additives		Measurements							Calculations		
Agents	Content	V_F	$t_{1/2}$	t_s	m_{50}	t_l	t_f	t_e	ρ_l	ρ_f	$V_G : V_L$
HV-CMC	0.6‰	610	10m43s	-	121.80	0s75	25s90	2s/40s	0.9720	0.1593	6.1
	0.8‰	600	12m10s	-	121.64	0s75	22s69	1s/13s	0.9688	0.1615	6.0
	1.0‰	620	12m32s	-	121.34	0s97	23s37	2s/15s	0.9628	0.1553	6.2
HEC	0.6‰	580	15m25s	-	123.39	0s94	28s94	2s/21s	1.0038	0.1731	5.8
	0.8‰	590	19m37s	-	122.02	0s90	34s34	2s/14s	0.9764	0.1655	5.9
	1.0‰	550	21m47s	-	122.36	0s78	40s47	1s/17s	0.9832	0.1788	5.5
HV-PAC	0.6‰	590	11m16s	-	121.98	0s84	22s02	2s/37s	0.9756	0.1654	5.9
	0.8‰	600	13m42s	-	122.31	0s82	35s06	2s/15s	0.9822	0.1637	6.0
	1.0‰	610	14m08s	-	122.39	0s91	23s69	2s/13s	0.9838	0.1613	6.1
Konjac	0.6‰	600	15m11s	-	124.30	0s81	46s90	1s/9s	1.0220	0.1703	6.0
	0.8‰	600	19m15s	1m53s	122.04	0s88	38s81	1s/7s	0.9768	0.1628	6.0
	1.0‰	600	21m24s	3m52s	123.33	0s78	52s81	1s/5s	1.0026	0.1671	6.0

Appendix C. 2: 5‰ K_{12} + 0.6‰ Sesbania Nr. 2

Additives		Measurements							Calculations		
Agents	Content	V_F	$t_{1/2}$	t_s	m_{50}	t_l	t_f	t_e	ρ_l	ρ_f	$V_G : V_L$
HV-CMC	0.6‰	620	10m02s	-	121.11	0s82	25s90	1s/18s	0.9582	0.1545	6.2
	0.8‰	620	11m59s	-	121.75	0s75	22s69	2s/14s	0.9710	0.1566	6.2
	1.0‰	660	11m15s	-	122.73	0s81	23s37	2s/27s	0.9906	0.1501	6.6
HEC	0.6‰	530	15m06s	-	121.61	0s75	28s94	2s/19s	0.9682	0.1827	5.3
	0.8‰	600	18m35s	-	121.80	0s97	34s34	2s/16s	0.9720	0.1620	6.0
	1.0‰	550	23m55s	-	122.37	0s88	40s47	2s/20s	0.9834	0.1788	5.5
HV-PAC	0.6‰	580	11m36s	-	121.54	0s87	22s02	2s/35s	0.9668	0.1667	5.8
	0.8‰	580	13m01s	-	123.71	0s81	35s06	2s/10s	1.0102	0.1742	5.8
	1.0‰	630	13m58s	-	121.39	0s72	23s69	2s/38s	0.9638	0.1530	6.3
Konjac	0.6‰	610	14m56s	-	123.47	0s72	46s90	1s/10s	1.0054	0.1648	6.1
	0.8‰	610	18m02s	1m47s	123.44	0s91	38s81	1s/7s	1.0048	0.1647	6.1
	1.0‰	620	20m24s	4m04s	122.58	1s06	52s81	1s/5s	0.9876	0.1593	6.2

Appendix C. 3: 5‰ K_{12} + 0.6‰ Sesbania Nr. 3

Additives		Measurements							Calculation		
Agents	Content	V_F	$t_{1/2}$	t_s	m_{50}	t_l	t_J	t_e	ρ_l	ρ_f	$V_G : V_L$
HV-CMC	0.6‰	650	8m21s	-	123.93	0s81	25s90	2s/16s	1.0146	0.1561	6.5
	0.8‰	630	14m41s	-	123.67	0s75	22s69	1s/13s	1.0094	0.1602	6.3
	1.0‰	620	12m17s	-	124.29	0s94	23s37	2s/29s	1.0218	0.1648	6.2
HEC	0.6‰	610	18m27s	-	122.39	0s72	28s94	2s/19s	0.9838	0.1613	6.1
	0.8‰	580	19m29s	-	123.72	0s79	34s34	2s/19s	1.0104	0.1742	5.8
	1.0‰	550	21m46s	-	120.79	0s97	40s47	2s/15s	0.9518	0.1731	5.5
HV-PAC	0.6‰	570	9m31s	-	123.42	0s75	22s02	2s/20s	1.0044	0.1762	5.7
	0.8‰	610	12m58s	-	120.99	0s84	35s06	1s/13s	0.9558	0.1567	6.1
	1.0‰	620	13m48s	-	123.72	0s66	23s69	2s/40s	1.0104	0.1630	6.2
Konjac	0.6‰	610	14m04s	-	120.84	0s87	46s90	1s/9s	0.9528	0.1562	6.1
	0.8‰	600	16m47s	1m44s	121.37	1s09	38s81	1s/7s	0.9634	0.1606	6.0
	1.0‰	610	21m55s	3m54s	121.09	0s91	52s81	1s/6s	0.9578	0.1570	6.1

Appendix C. 4: 5‰ K_{12} + 0.8‰ Sesbania Nr. 1

Additives		Measurements							Calculations		
Agents	Content	V_F	$t_{1/2}$	t_s	m_{50}	t_l	t_f	t_e	ρ_l	ρ_f	$V_G : V_L$
HV-CMC	0.6‰	630	13m18s	2m22s	121.71	0s88	31s53	2s/20s	0.9702	0.1540	6.3
	0.8‰	610	17m41s	2m28s	122.08	0s75	-	2s/15s	0.9776	0.1603	6.1
	1.0‰	600	18m02s	3m00s	122.34	0s69	32s84	1s/15s	0.9828	0.1638	6.0
HEC	0.6‰	590	15m27s	2m25s	121.62	0s81	55s35	2s/18s	0.9684	0.1641	5.9
	0.8‰	580	27m25s	3m25s	122.47	1s06	34s66	2s/11s	0.9854	0.1699	5.8
	1.0‰	560	22m27s	4m11s	122.03	0s85	40s12	2s/14s	0.9766	0.1744	5.6
HV-PAC	0.6‰	600	14m47s	2m41s	123.16	0s68	29s28	2s/34s	0.9992	0.1665	6.0
	0.8‰	560	15m54s	2m30s	122.52	0s97	26s60	1s/14s	0.9864	0.1761	5.6
	1.0‰	550	18m35s	2m44s	123.63	0s81	31s87	1s/12s	1.0086	0.1834	5.5
Konjac	0.6‰	610	16m17s	2m43s	121.73	0s81	36s50	2s/8s	0.9706	0.1591	6.1
	0.8‰	610	18m31s	3m13s	122.15	0s93	51s93	2s/10s	0.9790	0.1605	6.1
	1.0‰	640	22m50s	3m17s	123.52	0s90	1m06s	1s/6s	1.0064	0.1573	6.4

Appendix C. 5: 5‰ K_{12} + 0.8‰ Sesbania Nr. 2

Additives		Measurements							Calculations		
Agents	Content	V_F	$t_{1/2}$	t_s	m_{50}	t_l	t_f	t_e	ρ_l	ρ_f	$V_G : V_L$
HV-CMC	0.6‰	650	13m49s	1m50s	123.04	0s88	31s53	2s/14s	0.9968	0.1534	6.5
	0.8‰	620	16m22s	2m53s	123.02	0s81	?	2s/17s	0.9964	0.1607	6.2
	1.0‰	620	18m27s	3m08s	123.75	0s87	32s84	1s/15s	1.0110	0.1631	6.2
HEC	0.6‰	590	16m14s	2m38s	122.08	0s94	55s35	2s/18s	0.9776	0.1657	5.9
	0.8‰	600	26m11s	3m53s	123.62	0s87	34s66	2s/12s	1.0084	0.1681	6.0
	1.0‰	600	22m39s	3m23s	123.82	0s84	40s12	2s/13s	1.0124	0.1687	6.0
HV-PAC	0.6‰	610	16m05s	2m45s	122.28	0s75	29s28	1s/24s	0.9816	0.1609	6.1
	0.8‰	600	17m29s	2m28s	123.00	0s90	26s60	2s/10s	0.9960	0.1660	6.0
	1.0‰	600	18m58s	2m51s	121.61	0s84	31s87	1s/13s	0.9682	0.1614	6.0
Konjac	0.6‰	640	17m27s	2m36s	123.82	0s97	36s50	2s/9s	1.0124	0.1582	6.4
	0.8‰	590	18m39s	3m08s	123.37	0s94	51s93	1s/7s	1.0034	0.1701	5.9
	1.0‰	590	23m32s	3m40s	121.73	0s78	1m06s	1s/6s	0.9706	0.1645	5.9

Appendix C. 6: 5‰ K_{12} + 0.8‰ Sesbania Nr. 3

Additives		Measurements							Calculations		
Agents	Content	V_F	$t_{1/2}$	t_s	m_{50}	t_l	t_f	t_e	ρ_l	ρ_f	$V_G:V_L$
HV-CMC	0.6‰	610	12m30s	2m22s	121.79	0s75	31s53	1s/14s	0.9718	0.1593	6.1
	0.8‰	650	15m43s	2m20s	123.86	0s78	?	1s/14s	1.0132	0.1559	6.5
	1.0‰	610	18m59s	2m48s	121.81	0s75	32s84	2s/12s	0.9722	0.1594	6.1
HEC	0.6‰	580	16m22s	2m36s	123.27	1s06	55s35	1s/20s	1.0014	0.1727	5.8
	0.8‰	590	25m14s	4m09s	121.17	1s03	34s66	2s/11s	0.9594	0.1626	5.9
	1.0‰	610	23m38s	2m53s	122.19	0s91	40s12	2s/20s	0.9798	0.1606	6.1
HV-PAC	0.6‰	610	15m07s	2m36s	120.91	0s91	29s28	1s/21s	0.9542	0.1564	6.1
	0.8‰	600	17m57s	3m19s	121.28	0s90	26s60	1s/8s	0.9616	0.1603	6.0
	1.0‰	610	17m48s	3m25s	122.04	0s81	31s87	1s/12s	0.9768	0.1601	6.1
Konjac	0.6‰	640	18m14s	2m41s	122.54	0s94	36s50	2s/9s	0.9868	0.1549	6.4
	0.8‰	630	19m03s	3m24s	121.13	0s93	51s93	2s/7s	0.9586	0.1522	6.3
	1.0‰	630	21m55s	3m34s	121.93	0s88	1m06s	1s/6s	0.9746	0.1547	6.3

Appendix C. 7: 5‰ K_{12} + 1.0‰ Sesbania Nr.1

| Additives | | Measurements | | | | | | | Calculations | | |
Agents	Content	V_F	$t_{1/2}$	t_s	m_{50}	t_1	t_f	t_e	ρ_l	ρ_f	$V_G : V_L$
HV-CMC	0.6‰	620	18m33s	3m28s	123.98	0s88	27s28	2s/12s	1.0156	0.1638	6.2
	0.8‰	610	20m9s	3m04s	122.12	0s81	29s66	1s/10s	0.9784	0.1604	6.1
	1.0‰	600	19m48s	3m26s	123.15	0s99	28s22	1s/13s	0.9990	0.1665	6.0
HEC	0.6‰	610	25m15s	3m40s	123.69	0s84	1m03s	2s/12s	1.0098	0.1655	6.1
	0.8‰	600	25m22s	4m20s	122.25	0s87	35s50	2s/15s	0.9810	0.1635	6.0
	1.0‰	590	30m47s	4m48s	120.97	0s88	1m14s	2s/11s	0.9554	0.1619	5.9
HV-PAC	0.6‰	610	19m14s	3m07s	121.76	0s84	44s44	2s/9s	0.9712	0.1592	6.1
	0.8‰	600	20m31s	3m23s	121.67	0s75	33s37	1s/9s	0.9694	0.1616	6.0
	1.0‰	580	25m11s	3m50s	121.91	0s87	42s78	1s/10s	0.9742	0.1680	5.8
Konjac	0.6‰	630	19m52s	3m19s	123.57	0s91	44s91	1s/6s	1.0074	0.1599	6.3
	0.8‰	620	22m11s	4m07s	121.17	0s78	32s	2s/5s	0.9594	0.1547	6.2
	1.0‰	600	27m16s	4m34s	121.71	0s78	41s37	1s/4s	0.9702	0.1617	6.0

Appendix C. 8: 5‰ K_{12} + 1.0‰ Sesbania Nr.2

Additives		Measurements							Calculations		
Agents	Content	V_F	$t_{1/2}$	t_s	m_{50}	t_l	t_f	t_e	ρ_l	ρ_f	$V_G:V_L$
HV-CMC	0.6‰	620	17m26s	2m41s	122.92	0s79	27s28	1s/9s	0.9944	0.1604	6.2
	0.8‰	630	19m26s	3m23s	124.54	0s78	29s66	1s/12s	1.0268	0.1630	6.3
	1.0‰	620	19m36s	3m20s	124.01	0s75	28s22	1s/10s	1.0162	0.1639	6.2
HEC	0.6‰	610	22m03s	3m42s	121.97	1s03	1m03s	2s/10s	0.9754	0.1599	6.1
	0.8‰	610	23m22s	3m21s	122.83	1s15	35s50	2s/12s	0.9926	0.1627	6.1
	1.0‰	590	31m07s	6m11s	121.75	1s	1m14s	2s/10s	0.9710	0.1646	5.9
HV-PAC	0.6‰	630	18m05s	3m18s	122.73	0s94	44s44	1s/10s	0.9906	0.1572	6.3
	0.8‰	600	20m05s	3m29s	121.02	1s	33s37	2s/10s	0.9564	0.1594	6.0
	1.0‰	600	22m57s	3m41s	123.51	0s85	42s78	1s/8s	1.0062	0.1677	6.0
Konjac	0.6‰	620	17m48s	3m14s	123.27	0s81	44s91	1s/6s	1.0014	0.1615	6.2
	0.8‰	620	21m48s	3m32s	121.68	0s91	32s	2s/6s	0.9696	0.1564	6.2
	1.0‰	590	25m19s	3m43s	121.75	1s09	41s37	1s/4s	0.9710	0.1646	5.9

Appendix C. 9: 5‰ K_{12} + 1.0‰ Sesbania Nr.3

Additives		Measurements							Calculations		
Agents	Content	V_F	$t_{1/2}$	t_s	m_{50}	t_l	t_f	t_e	ρ_l	ρ_f	$V_G : V_L$
HV-CMC	0.6‰	630	18m19s	2m27s	121.96	0s84	27s28	1s/11s	0.9752	0.1548	6.3
	0.8‰	630	20m16s	3m36s	121.00	0s94	29s66	1s/11s	0.9560	0.1517	6.3
	1.0‰	610	23m37s	3m37s	122.09	0s84	28s22	1s/11s	0.9778	0.1603	6.1
HEC	0.6‰	580	21m57s	3m27s	121.19	1s	1m03s	2s/12s	0.9598	0.1655	5.8
	0.8‰	620	26m13s	4m03s	121.12	1s21	35s50	2s/12s	0.9584	0.1546	6.2
	1.0‰	610	31m14s	5m51s	122.77	0s90	1m14s	2s/10s	0.9914	0.1625	6.1
HV-PAC	0.6‰	590	18m23s	3m12s	123.53	0s84	44s44	1s/10s	1.0066	0.1706	5.9
	0.8‰	590	21m11s	3m47s	123.42	1s13	33s37	2s/10s	1.0044	0.1702	5.9
	1.0‰	610	22m44s	3m53s	122.89	0s82	42s78	1s/9s	0.9938	0.1629	6.1
Konjac	0.6‰	650	17m44s	2m55s	121.36	0s88	44s91	1s/7s	0.9632	0.1482	6.5
	0.8‰	630	23m12s	3m37s	122.69	1s03	32s	1s/6s	0.9898	0.1571	6.3
	1.0‰	590	26m17s	4m39s	123.48	1s09	41s37	1s/4s	1.0056	0.1704	5.9

Appendix C. 10: 4‰ ABS + 0.6‰ Sesbania Nr.1

Additives		Measurements							Calculations		
Agents	Content	V_F	$t_{1/2}$	t_s	m_{50}	t_l	t_f	t_e	ρ_l	ρ_f	$V_G:V_L$
HV-CMC	0.6‰	550	6m56s	-	123.39	0s84	0s75	2s	1.0038	0.1825	5.5
	0.8‰	480	6m55s	-	121.89	0s87	1s31	2s	0.9738	0.2029	4.8
	1.0‰	470	9m44s	-	121.26	0s78	-	1s	0.9612	0.2045	4.7
HEC	0.6‰	450	10m21s	-	122.41	0s78	-	1s	0.9842	0.2187	4.5
	0.8‰	460	11m50s	-	121.64	0s79	-	1s	0.9688	0.2106	4.6
	1.0‰	420	15m30s	-	123.38	0s82	-	1s	1.0036	0.2390	4.2
HV-PAC	0.6‰	450	9m15s	-	123.25	0s78	-	1s	1.0010	0.2224	4.5
	0.8‰	470	11m07s	-	121.13	0s78	-	1s	0.9586	0.2040	4.7
	1.0‰	450	12m53s	-	123.18	0s84	-	1s	0.9996	0.2221	4.5
Konjac	0.6‰	490	9m18s	-	121.49	0s85	-	1s	0.9658	0.1971	4.9
	0.8‰	470	-	-	122.89	0s87	-	1s	0.9938	0.2114	4.7
	1.0‰	450	15m53s	-	123.26	0s81	-	1s	1.0012	0.2225	4.5

Appendix C. 11: 4‰ ABS + 0.6‰ Sesbania Nr. 2

Additives		Measurements							Calculations		
Agents	Content	V_F	$t_{1/2}$	t_s	m_{50}	t_l	t_f	t_e	ρ_l	ρ_f	$V_G:V_L$
HV-CMC	0.6‰	500	5m48s	-	120.77	0s97	0s75	1s	0.9514	0.1903	5.0
	0.8‰	540	6m37s	-	123.07	0s84	1s31	1s	0.9974	0.1847	5.4
	1.0‰	480	9m19s	-	122.57	0s78	-	1s	0.9874	0.2057	4.8
HEC	0.6‰	490	9m43s	-	121.27	0s94	-	1s	0.9614	0.1962	4.9
	0.8‰	470	10m56s	-	121.61	0s81	-	1s	0.9682	0.2060	4.7
	1.0‰	450	15m23s	-	121.75	0s84	-	1s	0.9710	0.2158	4.5
HV-PAC	0.6‰	480	8m52s	-	121.03	0s87	-	1s	0.9566	0.1993	4.8
	0.8‰	500	10m50s	-	123.12	0s75	-	1s	0.9984	0.1997	5.0
	1.0‰	480	13m45s	-	121.05	0s79	-	1s	0.9570	0.1994	4.8
Konjac	0.6‰	480	9m52s	-	122.94	0s75	-	1s	0.9948	0.2073	4.8
	0.8‰	470	14m23s	-	121.73	0s75	-	1s	0.9706	0.2065	4.7
	1.0‰	460	15m24s	-	122.26	0s84	-	1s	0.9812	0.2133	4.6

Appendix C. 12: 4‰ ABS + 0.6‰ Sesbania Nr. 3

Additives		Measurements							Calculations		
Agents	Content	V_F	$t_{1/2}$	t_s	m_{50}	t_l	t_f	t_e	ρ_l	ρ_f	$V_G : V_L$
HV-CMC	0.6‰	510	6m16s	-	121.82	0s84	0s75	1s	0.9724	0.1907	5.1
	0.8‰	520	6m08s	-	120.80	0s88	1s31	1s	0.9520	0.1831	5.2
	1.0‰	490	9m28s	-	123.18	0s91	-	1s	0.9996	0.2040	4.9
HEC	0.6‰	470	10m41s	-	122.85	0s72	-	1s	0.9930	0.2113	4.7
	0.8‰	440	11m27s	-	123.31	0s85	-	1s	1.0022	0.2278	4.4
	1.0‰	430	14m39s	-	121.39	0s88	-	1s	0.9638	0.2241	4.3
HV-PAC	0.6‰	480	8m31s	-	122.30	0s75	-	1s	0.9820	0.2046	4.8
	0.8‰	500	11m32s	-	122.13	0s82	-	1s	0.9786	0.1957	5.0
	1.0‰	430	13m21s	-	121.52	0s78	-	1s	0.9664	0.2247	4.3
Konjac	0.6‰	500	9m19s	-	121.77	0s78	-	1s	0.9714	0.1943	5.0
	0.8‰	450	13m36s	-	121.93	0s78	-	1s	0.9746	0.2166	4.5
	1.0‰	440	15m57s	-	120.95	0s85	-	1s	0.9550	0.2170	4.4

Appendix C. 13: 4‰ ABS + 0.8‰ Sesbania Nr. 1

Additives		Measurements							Calculations		
Agents	Content	v_F	$t_{1/2}$	t_s	m_{50}	t_l	t_f	t_e	ρ_l	ρ_f	$v_G:v_L$
HV-CMC	0.6‰	440	9m37s	-	121.13	0s75	-	2s	0.9586	0.2179	4.4
	0.8‰	480	9m15s	-	122.13	0s84	-	2s	0.9786	0.2039	4.8
	1.0‰	460	10m04s	-	121.36	0s75	-	2s	0.9632	0.2094	4.6
HEC	0.6‰	440	12m05s	-	123.70	0s84	-	1s	1.0100	0.2295	4.4
	0.8‰	430	15m28s	-	121.19	0s75	-	1s	0.9598	0.2232	4.3
	1.0‰	470	14m48s	-	123.41	0s71	-	1s	1.0042	0.2137	4.7
HV-PAC	0.6‰	450	9m27s	-	122.26	0s78	-	2s	0.9812	0.2180	4.5
	0.8‰	420	12m17s	-	123.31	0s75	-	1s	1.0022	0.2386	4.2
	1.0‰	490	10m08s	-	123.50	0s85	-	1s	1.0060	0.2053	4.9
Konjac	0.6‰	460	13m03s	-	123.53	0s72	-	1s	1.0066	0.2188	4.6
	0.8‰	500	7m59s	-	121.49	0s84	2s35	1s	0.9658	0.1932	5.0
	1.0‰	420	18m19s	-	120.90	0s84	-	1s	0.9540	0.2271	4.2

Appendix C. 14: 4‰ ABS + 0.8‰ Sesbania Nr. 2

Additives		Measurements							Calculations		
Agents	Content	V_F	$t_{1/2}$	t_s	m_{50}	t_l	t_f	t_e	ρ_l	ρ_f	$V_G : V_L$
HV-CMC	0.6‰	480	9m28s	-	121.74	0s81	-	2s	0.9708	0.2023	4.8
	0.8‰	490	10m53s	-	123.96	0s81	-	2s	1.0152	0.2072	4.9
	1.0‰	440	12m40s	-	123.54	0s81	-	1s	1.0068	0.2288	4.4
HEC	0.6‰	470	11m33s	-	121.13	0s84	-	1s	0.9586	0.2040	4.7
	0.8‰	460	14m08s	-	123.03	0s82	-	1s	0.9966	0.2167	4.6
	1.0‰	460	15m51s	-	122.04	0s84	-	1s	0.9768	0.2123	4.6
HV-PAC	0.6‰	460	10m03s	-	122.06	0s85	-	1s	0.9772	0.2124	4.6
	0.8‰	450	12m43s	-	122.12	0s78	-	1s	0.9784	0.2174	4.5
	1.0‰	460	10m08s	-	121.79	0s75	-	1s	0.9718	0.2113	4.6
Konjac	0.6‰	480	11m50s	-	120.77	0s72	-	1s	0.9514	0.1982	4.8
	0.8‰	520	8m48s	-	122.73	0s81	2s35	1s	0.9906	0.1905	5.2
	1.0‰	440	18m35s	-	122.84	0s94	-	1s	0.9928	0.2256	4.4

Appendix C. 15: 4‰ ABS + 0.8‰ Sesbania Nr. 3

Additives		Measurements							Calculations		
Agents	Content	V_F	$t_{1/2}$	t_s	m_{50}	t_l	l_f	t_e	ρ_l	ρ_f	$V_G : V_L$
HV-CMC	0.6‰	470	10m21s	-	123.80	0s82	-	2s	1.0120	0.2153	4.7
	0.8‰	440	13m37s	-	121.93	0s84	-	2s	0.9746	0.2215	4.4
	1.0‰	450	18m26s	-	122.37	0s90	-	2s	0.9834	0.2185	4.5
HEC	0.6‰	490	12m39s	-	121.96	0s81	-	1s	0.9752	0.1990	4.9
	0.8‰	460	13m47s	-	121.35	0s87	-	1s	0.9630	0.2093	4.6
	1.0‰	470	16m34s	-	122.15	0s91	-	1s	0.9790	0.2083	4.7
HV-PAC	0.6‰	460	9m37s	-	123.73	0s75	-	1s	1.0106	0.2197	4.6
	0.8‰	450	12m11s	-	120.98	0s75	-	1s	0.9556	0.2124	4.5
	1.0‰	450	10m18s	-	120.42	0s84	-	2s	0.9444	0.2099	4.5
Konjac	0.6‰	480	10m48s	-	122.04	0s81	-	1s	0.9768	0.2035	4.8
	0.8‰	510	8m38s	-	123.44	0s85	2s35	1s	1.0048	0.1970	5.1
	1.0‰	420	17m38s	-	121.60	0s91	-	1s	0.9680	0.2305	4.2

Appendix C. 16: 4‰ ABS + 1.0‰ Sesbania Nr. 1

Additives		Measurements							Calculations		
Agents	Content	V_F	$t_{1/2}$	t_s	m_{50}	t_l	l_f	t_e	ρ_l	ρ_f	$V_G:V_L$
HV-CMC	0.6‰	470	12m04s	-	121.25	0s78	-	2s	0.9610	0.2045	4.7
	0.8‰	440	10m49s	-	123.17	0s87	-	1s	0.9994	0.2271	4.4
	1.0‰	430	12m03s	-	122.30	0s75	-	2	0.9820	0.2284	4.3
HEC	0.6‰	410	13m43s	-	120.99	0s90	-	1s	0.9558	0.2331	4.1
	0.8‰	410	18m18s	-	120.75	0s81	-	1s	0.9510	0.2320	4.1
	1.0‰	400	20m05s	-	121.22	1s	-	2s	0.9604	0.2401	4.0
HV-PAC	0.6‰	420	11m57s	-	121.85	0s82	-	1s	0.9730	0.2317	4.2
	0.8‰	450	11m55s	-	121.94	0s82	-	1s	0.9748	0.2166	4.5
	1.0‰	400	16m01s	-	121.92	0s75	-	1s	0.9744	0.2436	4.0
Konjac	0.6‰	420	15m24s	-	122.11	0s82	-	2s	0.9782	0.2329	4.2
	0.8‰	440	18m34s	-	121.91	0s78	-	2s	0.9742	0.2214	4.4
	1.0‰	390	21m42s	-	121.72	0s88	-	2s	0.9704	0.2488	3.9

Appendix C. 17: 4‰ ABS + 1.0‰ Sesbania Nr. 2

Additives		Measurements							Calculations		
Agents	Content	V_F	$t_{1/2}$	t_s	m_{50}	t_l	t_f	t_e	ρ_l	ρ_f	$V_G : V_L$
HV-CMC	0.6‰	470	11m37s	-	123.44	0s82	-	2s	1.0048	0.2138	4.7
	0.8‰	450	12m01s	-	122.51	0s75	-	2s	0.9862	0.2192	4.5
	1.0‰	440	11m57s	-	123.33	0s84	-	2s	1.0026	0.2279	4.4
HEC	0.6‰	410	15m15s	-	123.07	0s87	-	1s	0.9974	0.2433	4.1
	0.8‰	420	19m06s	-	123.23	0s83	-	1s	1.0006	0.2382	4.2
	1.0‰	400	20m15s	-	122.68	0s81	-	1s	0.9896	0.2474	4.0
HV-PAC	0.6‰	440	12m43s	-	122.47	0s94	-	2s	0.9854	0.2240	4.4
	0.8‰	440	12m43s	-	123.52	0s78	-	2s	1.0064	0.2287	4.4
	1.0‰	420	15m51s	-	121.05	0s87	-	2s	0.9570	0.2279	4.2
Konjac	0.6‰	440	14m18s	-	123.80	1s	-	2s	1.0120	0.2300	4.4
	0.8‰	440	17m51s	-	123.65	0s90	-	2s	1.0090	0.2293	4.4
	1.0‰	410	22m15s	-	124.21	0s91	-	2s	1.0202	0.2488	4.1

Appendix C. 18: 4‰ ABS + 1.0‰ Sesbania Nr. 3

Additives		Measurements							Calculations		
Agents	Content	V_F	$t_{1/2}$	t_s	m_{50}	t_l	t_f	t_e	ρ_l	ρ_f	$V_G : V_L$
HV-CMC	0.6‰	460	11m44s	-	121.92	0s75	-	2s	0.9744	0.2118	4.6
	0.8‰	440	14m06s	-	122.56	?	-	2s	0.9872	0.2244	4.4
	1.0‰	420	15m52s	-	120.65	0s69	-	2s	0.9490	0.2260	4.2
HEC	0.6‰	450	13m32s	-	122.00	1s06	-	1s	0.9760	0.2169	4.5
	0.8‰	430	17m14s	-	122.09	0s91	-	1s	0.9778	0.2274	4.3
	1.0‰	410	19m46s	-	121.27	1s	-	1s	0.9614	0.2345	4.1
HV-PAC	0.6‰	430	11m38s	-	123.73	0s91	-	1s	1.0106	0.2350	4.3
	0.8‰	440	13m13s	-	122.41	0s91	-	1s	0.9842	0.2237	4.4
	1.0‰	420	18m51s	-	123.53	0s81	-	2s	1.0066	0.2397	4.2
Konjac	0.6‰	420	14m09s	-	121.35	0s78	-	2s	0.9630	0.2293	4.2
	0.8‰	400	20m16s	-	121.02	0s78	-	2s	0.9564	0.2391	4.0
	1.0‰	410	24m44s	-	122.86	0s78	-	2s	0.9932	0.2422	4.1